CONTENTS

PART I **Principles**

PART II **Methods**

PART III **Forms**

PART IV **Attitudes**

BOOK 1

PART I

Principles

1

The Transaction

A school in Connecticut once held "a day devoted to the arts," and I was asked if I would come and talk about writing as a vocation. When I arrived I found that a second speaker had been invited—Dr. Brock (as I'll call him), a surgeon who had recently begun to write and had sold some stories to magazines. He was going to talk about writing as an avocation. That made us a panel, and we sat down to face a crowd of students and teachers and parents, all eager to learn the secrets of our glamorous work.

Dr. Brock was dressed in a bright red jacket, looking vaguely bohemian, as authors are supposed to look, and the first question went to him. What was it like to be a writer?

He said it was tremendous fun. Coming home from an arduous day at the hospital, he would go straight to his yellow pad and write his tensions away. The words just flowed. It was easy. I then said that writing wasn't easy and wasn't fun. It was hard and lonely, and the words seldom just flowed.

Next Dr. Brock was asked if it was important to rewrite. Absolutely not, he said. "Let it all hang out," he told us, and whatever form the sentences take will reflect the writer at his most natural. I then said that rewriting is the essence of writing. I pointed out that professional writers rewrite their sentences over and over and then rewrite what they have rewritten.

"What do you do on days when it isn't going well?" Dr. Brock was asked. He said he just stopped writing and put the work aside for a day when it would go better. I then said that the professional writer must establish a daily schedule and stick to it. I said that writing is a craft, not an art, and that the man who runs away from his craft because he lacks inspiration is fooling himself. He is also going broke.

"What if you're feeling depressed or unhappy?" a student asked. "Won't that affect your writing?"

Probably it will, Dr. Brock replied. Go fishing. Take a walk. Probably it won't, I said. If your job is to write every day, you learn to do it like any other job.

A student asked if we found it useful to circulate in the literary world. Dr. Brock said he was greatly enjoying his new life as a man of letters, and he told several stories of being taken to lunch by his publisher and his agent at Manhattan restaurants where writers and editors gather. I said that professional writers are solitary drudges who seldom see other writers.

"Do you put symbolism in your writing?" a student asked me.

"Not if I can help it," I replied. I have an unbroken record of missing the deeper meaning in any story, play or movie, and as for dance and mime, I have never had any idea of what is being conveyed.

"I *love* symbols!" Dr. Brock exclaimed, and he described with gusto the joys of weaving them through his work.

So the morning went, and it was a revelation to all of us. At the end Dr. Brock told me he was enormously interested in my answers—it had never occurred to him that writing could be hard. I told him I was just as interested in *his* answers —it had never occurred to me that writing could be easy. Maybe I should take up surgery on the side.

As for the students, anyone might think we left them bewildered. But in fact we gave them a broader glimpse of the writing process than if only one of us had talked. For there isn't any "right" way to do such personal work. There are all kinds of writers and all kinds of methods, and any method that helps you to say what you want to say is the right method for you. Some people write by day, others by night. Some people need silence, others turn on the radio. Some write by hand, some by computer, some by talking into a tape recorder. Some people write their first draft in one long burst and then revise; others can't write the second paragraph until they have fiddled endlessly with the first.

But all of them are vulnerable and all of them are tense. They are driven by a compulsion to put some part of themselves on paper, and yet they don't just write what comes naturally. They sit down to commit an act of literature, and the self who emerges on paper is far stiffer than the person who sat down to write. The problem is to find the real man or woman behind the tension.

Ultimately the product that any writer has to sell is not the subject being written about, but who he or she is. I often find myself reading with interest about a topic I never thought would interest me—some scientific quest, perhaps. What holds me is the enthusiasm of the writer for his field. How was he drawn into it? What emotional baggage did he bring along? How did it change his life? It's not necessary to want to spend a year alone at Walden Pond to become involved with a writer who did.

This is the personal transaction that's at the heart of good nonfiction writing. Out of it come two of the most important qualities that this book will go in search of: humanity and warmth. Good writing has an aliveness that keeps the reader reading from one paragraph to the next, and it's not a question of gimmicks to "personalize" the author. It's a question of using the English language in a way that will achieve the greatest clarity and strength.

Can such principles be taught? Maybe not. But most of them can be learned.

2

<u>Simplicity</u>

Clutter is the disease of American writing. We are a society strangling in unnecessary words, circular constructions, pompous frills and meaningless jargon.

Who can understand the clotted language of everyday American commerce: the memo, the corporation report, the business letter, the notice from the bank explaining its latest "simplified" statement? What member of an insurance or medical plan can decipher the brochure explaining his costs and benefits? What father or mother can put together a child's toy from the instructions on the box? Our national tendency is to inflate and thereby sound important. The airline pilot who announces that he is presently anticipating experiencing considerable precipitation wouldn't think of saying it may rain. The sentence is too simple— there must be something wrong with it.

But the secret of good writing is to strip every sentence to its cleanest components. Every word that serves no function, every long word that could be a short word, every adverb that carries the same meaning that's already in the verb, every passive construction that leaves the reader unsure of who is doing what—these are the thousand and one adulterants that weaken the strength of a sentence. And they usually occur in proportion to education and rank.

During the 1960s the president of my university wrote a letter to mollify the alumni after a spell of campus unrest. "You are probably aware," he began, "that we have been experiencing very considerable potentially explosive expressions of dissatisfaction on issues only partially related." He meant that the students had been hassling them about different things. I was far more upset by the president's English than by the students' potentially explosive expressions of dissatisfaction. I would have preferred the presidential approach taken by Franklin D. Roosevelt when he tried to convert into English his own government's memos, such as this blackout order of 1942:

Such preparations shall be made as will completely obscure all Federal buildings and non-Federal buildings occupied by the Federal government during an air raid for any period of time from visibility by reason of internal or external illumination.

"Tell them," Roosevelt said, "that in buildings where they have to keep the work going to put something across the windows."

Simplify, simplify. Thoreau said it, as we are so often reminded, and no American writer more consistently practiced what he preached. Open *Walden* to any page and you will find a man saying in a plain and orderly way what is on his mind:

I went to the woods because I wished to live deliberately, to front only the essential facts of life, and see if I could not learn what it had to teach, and not, when I came to die, discover that I had not lived.

How can the rest of us achieve such enviable freedom from clutter? The answer is to clear our heads of clutter. Clear thinking becomes clear writing; one can't exist without the other. It's impossible for a muddy thinker to write good English. He may get away with it for a paragraph or two, but soon the reader will be lost, and there's no sin so grave, for the reader will not easily be lured back.

Who is this elusive creature, the reader? The reader is someone with an attention span of about 30 seconds—a person assailed by many forces competing for attention. At one time those forces were relatively few: newspapers, magazines, radio, spouse, children, pets. Today they also include a galaxy of electronic devices for receiving entertainment and information—television, VCRs, DVDs, CDs, video games, the Internet, e-mail, cell phones, BlackBerries, iPods—as well as a fitness program, a pool, a lawn and that most potent of competitors, sleep. The man or woman snoozing in a chair with a magazine or a book is a person who was being given too much unnecessary trouble by the writer.

It won't do to say that the reader is too dumb or too lazy to keep pace with the train of thought. If the reader is lost, it's usually because the writer hasn't been careful enough. That carelessness can take any number of forms. Perhaps a sentence is so excessively cluttered that the reader, hacking through the verbiage, simply doesn't know what it means. Perhaps a sentence has been so shoddily constructed that the reader could read it in several ways. Perhaps the writer has switched pronouns in midsentence, or has switched tenses, so the reader loses track of who is talking or when the action took place. Perhaps

Sentence B is not a logical sequel to Sentence A; the writer, in whose head the connection is clear, hasn't bothered to provide the missing link. Perhaps the writer has used a word incorrectly by not taking the trouble to look it up.

Faced with such obstacles, readers are at first tenacious. They blame themselves—they obviously missed something, and they go back over the mystifying sentence, or over the whole paragraph, piecing it out like an ancient rune, making guesses and moving on. But they won't do that for long. The writer is making them work too hard, and they will look for one who is better at the craft.

Writers must therefore constantly ask: what am I trying to say? Surprisingly often they don't know. Then they must look at what they have written and ask: have I said it? Is it clear to someone encountering the subject for the first time? If it's not, some fuzz has worked its way into the machinery. The clear writer is someone clearheaded enough to see this stuff for what it is: fuzz.

I don't mean that some people are born clearheaded and are therefore natural writers, whereas others are naturally fuzzy and will never write well. Thinking clearly is a conscious act that writers must force on themselves, as if they were working on any other project that requires logic: making a shopping list or doing an algebra problem. Good writing doesn't come naturally, though most people seem to think it does. Professional writers are constantly bearded by people who say they'd like to "try a little writing sometime"—meaning when they retire from their real profession, like insurance or real estate, which is hard. Or they say, "I could write a book about that." I doubt it.

Writing is hard work. A clear sentence is no accident. Very few sentences come out right the first time, or even the third time. Remember this in moments of despair. If you find that writing is hard, it's because it *is* hard.

is too dumb or too lazy to keep pace with the ~~writer's~~ train of thought. My sympathies are ~~entirely~~ with him.) ~~He's not so dumb.~~ (If the reader is lost, it is generally because the writer ~~of the article~~ has not been careful enough to keep him on the ~~proper~~ path.

This carelessness can take any number of ~~different~~ forms. Perhaps a sentence is so excessively ~~long and~~ cluttered that the reader, hacking his way through ~~all~~ the verbiage, simply doesn't know what [it] ~~the writer~~ means. Perhaps a sentence has been so shoddily constructed that the reader could read it in any of [several] ~~two or three different~~ ways. ~~He thinks he knows what the writer is trying to say, but he's not sure.~~ Perhaps the writer has switched pronouns in mid-sentence, or ~~perhaps he~~ has switched tenses, so the reader loses track of who is talking ~~to whom,~~ or ~~exactly~~ when the action took place. Perhaps Sentence B is not a logical sequel to Sentence A -- the writer, in whose head the connection is ~~perfectly~~ clear, has not [bothered to provide] ~~given enough thought to providing~~ the missing link. Perhaps the writer has used an important word incorrectly by not taking the trouble to look it up ~~and make sure.~~ He may think that "sanguine" and "sanguinary" mean the same thing, but) ~~I can assure you that~~ (the difference is a bloody big one ~~to the reader.~~ [The reader] ~~He~~ can only ~~try to~~ infer ~~what~~ (speaking of big differences) what the writer is trying to imply.

Faced with [mere] ~~such a variety of~~ obstacles, the reader is at first a remarkably tenacious bird. He ~~tends to~~ blame[s] himself[.] ~~He~~ obviously missed something, ~~he thinks,~~ and he goes back over the mystifying sentence, or over the whole paragraph,

piecing it out like an ancient rune, making guesses and moving
on. But he won't do this for long. ~~He will soon run out of
patience.~~ (The writer is making him work too hard ~~—harder
than he should have to work~~ —(and the reader will look for
~~a writer~~ [one] who is better at his craft.

The writer must therefore constantly ask himself: What am
I trying to say? ~~in this sentence?~~ (Surprisingly often, he
doesn't know. ~~And~~ [T]hen he must look at what he has ~~just~~
written and ask: Have I said it? Is it clear to someone
[encountering] ~~who is coming upon~~ the subject for the first time? If it's
not, ~~clear,~~ it is because some fuzz has worked its way into the
machinery. The clear writer is a person ~~who is~~ clear-headed
enough to see this stuff for what it is: fuzz.

I don't mean ~~to suggest~~ that some people are born
clear-headed and are therefore natural writers, whereas
[others] ~~other people~~ are naturally fuzzy and will ~~therefore~~ never write
well. Thinking clearly is ~~an entirely~~ conscious act that the
writer must [force] ~~keep forcing~~ upon himself, just as if he were
[embarking] ~~starting out~~ on any other ~~kind of~~ project that ~~calls for~~ [requires] logic:
adding up a laundry list or doing an algebra problem ~~or playing
chess.~~ Good writing doesn't ~~just~~ come naturally, though most
people obviously think [it does.] ~~it's as easy as walking.~~ The professional

Two pages of the final manuscript of this chapter from the First Edition of *On Writing Well*. Although they look like a first draft, they had already been rewritten and retyped—like almost every other page—four or five times. With each rewrite I try to make what I have written tighter, stronger and more precise, eliminating every element that's not doing useful work. Then I go over it once more, reading it aloud, and am always amazed at how much clutter can still be cut. (In later editions I eliminated the sexist pronoun "he" denoting "the writer" and "the reader.")

<u>**3**</u>

⸺ᴏⱱᴏ⸺

<u>Clutter</u>

Fighting clutter is like fighting weeds—the writer is always slightly behind. New varieties sprout overnight, and by noon they are part of American speech. Consider what President Nixon's aide John Dean accomplished in just one day of testimony on television during the Watergate hearings. The next day everyone in America was saying "at this point in time" instead of "now."

Consider all the prepositions that are draped onto verbs that don't need any help. We no longer head committees. We head them up. We don't face problems anymore. We face up to them when we can free up a few minutes. A small detail, you may say—not worth bothering about. It *is* worth bothering about. Writing improves in direct ratio to the number of things we can keep out of it that shouldn't be there. "Up" in "free up" shouldn't be there. Examine every word you put on paper. You'll find a surprising number that don't serve any purpose.

Take the adjective "personal," as in "a personal friend of mine," "his personal feeling" or "her personal physician." It's typical of hundreds of words that can be eliminated. The personal friend has come into the language to distinguish him or her from the business friend, thereby debasing both language and friendship. Someone's feeling *is* that person's personal feeling—that's what "his" means. As for the personal physician, that's the man or woman summoned to the dressing room of a stricken actress so she won't have to be treated by the impersonal physician assigned to the theater. Someday I'd like to see that person identified as "her doctor." Physicians are physicians, friends are friends. The rest is clutter.

Clutter is the laborious phrase that has pushed out the short word that means the same thing. Even before John Dean, people and businesses had stopped saying "now." They were saying "currently" ("all our operators are currently assisting other customers"), or "at the present time," or "presently" (which means "soon"). Yet the idea can always be expressed by "now" to mean the immediate moment ("Now I can see him"), or by "today" to mean the historical present ("Today prices are high"), or simply by the verb "to be" ("It is raining"). There's no need to say, "At the present time we are experiencing precipitation."

"Experiencing" is one of the worst clutterers. Even your dentist will ask if you are experiencing any pain. If he had his own kid in the chair he would say, "Does it hurt?" He would, in short, be himself. By using a more pompous phrase in his professional role he not only sounds more important; he blunts the painful edge of truth. It's the language of the flight attendant demonstrating the oxygen mask that will drop down if the plane should run out of air. "In the unlikely possibility that the aircraft should experience such an eventuality," she begins— a phrase so oxygen-depriving in itself that we are prepared for any disaster.

Clutter is the ponderous euphemism that turns a slum into a depressed socioeconomic area, garbage collectors into waste-disposal personnel and the town dump into the volume reduction unit. I think of Bill Mauldin's cartoon of two hoboes riding a freight car. One of them says, "I started as a simple bum, but now I'm hard-core unemployed." Clutter is political correctness gone amok. I saw an ad for a boys' camp designed to provide "individual attention for the minimally exceptional."

Clutter is the official language used by corporations to hide their mistakes. When the Digital Equipment Corporation eliminated 3,000 jobs its statement didn't mention layoffs; those were "involuntary methodologies." When an Air Force missile crashed, it "impacted with the ground prematurely." When General Motors had a plant shutdown, that was a "volume-related production-schedule adjustment." Companies that go belly-up have "a negative cash-flow position."

Clutter is the language of the Pentagon calling an invasion a "reinforced protective reaction strike" and justifying its vast budgets on the need for "counterforce deterrence." As George Orwell pointed out in "Politics and the English Language," an essay written in 1946 but often cited during the wars in Cambodia, Vietnam and Iraq, "political speech and writing are largely the defense of the indefensible.... Thus political language has to consist largely of euphemism, question-begging and sheer cloudy vagueness." Orwell's warning that clutter is not just a nuisance but a deadly tool has come true in the recent decades of American military adventurism. It was during George W. Bush's presidency that "civilian casualties" in Iraq became "collateral damage."

Verbal camouflage reached new heights during General Alexander Haig's tenure as President Reagan's secretary of state. Before Haig nobody had thought of saying "at this juncture of maturization" to mean "now." He told the American people that terrorism could be fought with "meaningful sanctionary teeth" and that intermediate nuclear missiles were "at the vortex of cruciality." As for any worries that the public might harbor, his message was "leave it to Al," though what he actually said was: "We must push this to a lower decibel of public fixation. I don't think there's much of a learning curve to be achieved in this area of content."

I could go on quoting examples from various fields—every profession has its growing arsenal of jargon to throw dust in the eyes of the populace. But the list would be tedious. The point of raising it now is to serve notice that clutter is the enemy. Beware, then, of the long word that's no better than the short word: "assistance" (help), "numerous" (many), "facilitate" (ease), "individual" (man or woman), "remainder" (rest), "initial" (first), "implement" (do), "sufficient" (enough), "attempt" (try), "referred to as" (called) and hundreds more. Beware of all the slippery new fad words: paradigm and parameter, prioritize and potentialize. They are all weeds that will smother what you write. Don't dialogue with someone you can talk to. Don't interface with anybody.

Just as insidious are all the word clusters with which we explain how we propose to go about our explaining: "I might add," "It should be pointed out," "It is interesting to note." If you might add, add it. If it should be pointed out, point it out. If it is interesting to note, *make* it interesting; are we not all stupefied by what follows when someone says, "This will interest you"? Don't inflate what needs no inflating: "with the possible exception of" (except), "due to the fact that" (because), "he totally lacked the ability to" (he couldn't), "until such time as" (until), "for the purpose of" (for).

Is there any way to recognize clutter at a glance? Here's a device my students at Yale found helpful. I would put brackets around every component in a piece of writing that wasn't doing useful work. Often just one word got bracketed: the unnecessary preposition appended to a verb ("order up"), or the adverb that carries the same meaning as the verb ("smile happily"), or the adjective that states a known fact ("tall skyscraper"). Often my brackets surrounded the little qualifiers that weaken any sentence they inhabit ("a bit," "sort of"), or phrases like "in a sense," which don't mean anything. Sometimes my brackets surrounded an entire sentence—the one that essentially repeats what the previous sentence said, or that says something readers don't need to know or can figure out for themselves. Most first drafts can be cut by 50 percent without losing any information or losing the author's voice.

My reason for bracketing the students' superfluous words, instead of crossing them out, was to avoid violating their sacred prose. I wanted to leave the sentence intact for them to analyze. I was saying, "I may be wrong, but I think this can be deleted and the meaning won't be affected. But *you* decide. Read the sentence without the bracketed material and see if it works." In the early weeks of the term I handed back papers that were festooned with brackets. Entire paragraphs were bracketed. But soon the students learned to put mental brackets around their own clutter, and by the end of the term their papers were almost clean. Today many of those students are professional writers, and they tell me, "I still see your brackets—they're following me through life."

You can develop the same eye. Look for the clutter in your writing and prune it ruthlessly. Be grateful for everything you can throw away. Reexamine each sentence you put on paper. Is every word doing new work? Can any thought be expressed with more economy? Is anything pompous or pretentious or faddish? Are you hanging on to something useless just because you think it's beautiful?

Simplify, simplify.

<u>4</u>

—⌇⌇⌇—

<u>Style</u>

So much for early warnings about the bloated monsters that lie in ambush for the writer trying to put together a clean English sentence.

"But," you may say, "if I eliminate everything you think is clutter and if I strip every sentence to its barest bones, will there be anything left of me?" The question is a fair one; simplicity carried to an extreme might seem to point to a style little more sophisticated than "Dick likes Jane" and "See Spot run."

I'll answer the question first on the level of carpentry. Then I'll get to the larger issue of who the writer is and how to preserve his or her identity.

Few people realize how badly they write. Nobody has shown them how much excess or murkiness has crept into their style and how it obstructs what they are trying to say. If you give me an eight-page article and I tell you to cut it to four pages, you'll howl and say it can't be done. Then you'll go home and do it, and it will be much better. After that comes the hard part: cutting it to three.

The point is that you have to strip your writing down before you can build it back up. You must know what the essential tools are and what job they were designed to do. Extending the metaphor of carpentry, it's first necessary to be able to saw wood neatly and to drive nails. Later you can bevel the edges or add elegant finials, if that's your taste. But you can never forget that you are practicing a craft that's based on certain principles. If the nails are weak, your house will collapse. If your verbs are weak and your syntax is rickety, your sentences will fall apart.

I'll admit that certain nonfiction writers, like Tom Wolfe and Norman Mailer, have built some remarkable houses. But these are writers who spent years learning their craft, and when at last they raised their fanciful turrets and hanging gardens, to the surprise of all of us who never dreamed of such ornamentation, they knew what they were doing. Nobody becomes Tom Wolfe overnight, not even Tom Wolfe.

First, then, learn to hammer the nails, and if what you build is sturdy and serviceable, take satisfaction in its plain strength.

But you will be impatient to find a "style"—to embellish the plain words so that readers will recognize you as someone special. You will reach for gaudy similes and tinseled adjectives, as if "style" were something you could buy at the style store and drape onto your words in bright decorator colors. (Decorator colors are the colors that decorators come in.) There is no style store; style is organic to the person doing the writing, as much a part of him as his hair, or, if he is bald, his lack of it. Trying to add style is like adding a toupee. At first glance the formerly bald man looks young and even handsome. But at second glance—and with a toupee there's always a second glance—he doesn't look quite right. The problem is not that he doesn't look well groomed; he does, and we can only admire the wigmaker's skill. The point is that he doesn't look like himself.

This is the problem of writers who set out deliberately to garnish their prose. You lose whatever it is that makes you unique. The reader will notice if you are putting on airs. Readers want the person who is talking to them to sound genuine. Therefore a fundamental rule is: be yourself.

No rule, however, is harder to follow. It requires writers to do two things that by their metabolism are impossible. They must relax, and they must have confidence.

Telling a writer to relax is like telling a man to relax while being examined for a hernia, and as for confidence, see how stiffly he sits, glaring at the screen that awaits his words. See how often he gets up to look for something to eat or drink. A writer will do anything to avoid the act of writing. I can testify from my newspaper days that the number of trips to the water cooler per reporter-hour far exceeds the body's need for fluids.

What can be done to put the writer out of these miseries? Unfortunately, no cure has been found. I can only offer the consoling thought that you are not alone. Some days will go better than others. Some will go so badly that you'll despair of ever writing again. We have all had many of those days and will have many more.

Still, it would be nice to keep the bad days to a minimum, which brings me back to the problem of trying to relax.

Assume that you are the writer sitting down to write. You think your article must be of a certain length or it won't seem important. You think how august it will look in print. You think of all the people who will read it. You think that it must have the solid weight of authority. You think that its style must dazzle. No wonder you tighten; you are so busy thinking of your awesome responsibility to the finished article that you can't even start. Yet you vow to be worthy of the task, and, casting about for grand phrases that wouldn't occur to you if you weren't trying so hard to make an impression, you plunge in.

Paragraph 1 is a disaster—a tissue of generalities that seem to have come out of a machine. No *person* could have written them. Paragraph 2 isn't much better. But Paragraph 3 begins to have a somewhat human quality, and by Paragraph 4 you begin to sound like yourself. You've started to relax. It's amazing how often an editor can throw away the first three or four paragraphs of an article, or even the first few pages, and start with the paragraph where the writer begins to sound like himself or herself. Not only are those first paragraphs impersonal and ornate; they don't say anything—they are a self-conscious attempt at a fancy prologue. What I'm always looking for as an editor is a sentence that says something like "I'll never forget the day when I …" I think, "Aha! A person!"

Writers are obviously at their most natural when they write in the first person. Writing is an intimate transaction between two people, conducted on paper, and it will go well to the extent that it retains its humanity. Therefore I urge people to write in the first person: to use "I" and "me" and "we" and "us." They put up a fight.

"Who am I to say what *I* think?" they ask. "Or what *I* feel?"

"Who are you *not* to say what you think?" I tell them. "There's only one you. Nobody else thinks or feels in exactly the same way."

"But nobody cares about my opinions," they say. "It would make me feel conspicuous."

"They'll care if you tell them something interesting," I say, "and tell them in words that come naturally."

Nevertheless, getting writers to use "I" is seldom easy. They think they must earn the right to reveal their emotions or their thoughts. Or that it's egotistical. Or that it's undignified—a fear that afflicts the academic world. Hence the professorial use of "one" ("One finds oneself not wholly in accord with Dr. Maltby's view of the human condition"), or of the impersonal "it is" ("It is to be hoped that Professor Felt's monograph will find the wider audience it most assuredly deserves"). I don't want to meet "one"—he's a boring guy. I want a professor with a passion for his subject to tell me why it fascinates *him*.

I realize that there are vast regions of writing where "I" isn't allowed. Newspapers don't want "I" in their news stories; many magazines don't want it in their articles; businesses and institutions don't want it in the reports they send so profusely into the American home; colleges don't want "I" in their term papers or dissertations, and English teachers discourage any first-person pronoun except the literary "we" ("We see in Melville's symbolic use of the white whale

...”). Many of those prohibitions are valid; newspaper articles should consist of news, reported objectively. I also sympathize with teachers who don't want to give students an easy escape into opinion—"I think Hamlet was stupid"—before they have grappled with the discipline of assessing a work on its merits and on external sources. "I" can be a self-indulgence and a cop-out.

Still, we have become a society fearful of revealing who we are. The institutions that seek our support by sending us their brochures sound remarkably alike, though surely all of them—hospitals, schools, libraries, museums, zoos—were founded and are still sustained by men and women with different dreams and visions. Where are these people? It's hard to glimpse them among all the impersonal passive sentences that say "initiatives were undertaken" and "priorities have been identified."

Even when "I" isn't permitted, it's still possible to convey a sense of I-ness. The political columnist James Reston didn't use "I" in his columns; yet I had a good idea of what kind of person he was, and I could say the same of many other essayists and reporters. Good writers are visible just behind their words. If you aren't allowed to use "I," at least think "I" while you write, or write the first draft in the first person and then take the "I"s out. It will warm up your impersonal style.

Style is tied to the psyche, and writing has deep psychological roots. The reasons why we express ourselves as we do, or fail to express ourselves because of "writer's block," are partly buried in the subconscious mind. There are as many kinds of writer's block as there are kinds of writers, and I have no intention of trying to untangle them. This is a short book, and my name isn't Sigmund Freud.

But I've also noticed a new reason for avoiding "I": Americans are unwilling to go out on a limb. A generation ago our leaders told us where they stood and what they believed. Today they perform strenuous verbal feats to escape that fate. Watch them wriggle through TV interviews without committing themselves. I remember President Ford assuring a group of visiting businessmen that his fiscal policies would work. He said: "We see nothing but increasingly brighter clouds every month." I took this to mean that the clouds were still fairly dark. Ford's sentence was just vague enough to say nothing and still sedate his constituents.

Later administrations brought no relief. Defense Secretary Caspar Weinberger, assessing a Polish crisis in 1984, said: "There's continuing ground for serious concern and the situation remains serious. The longer it remains serious, the more ground there is for serious concern." The first President Bush, questioned about his stand on assault rifles, said: "There are various groups that think you can ban certain kinds of guns. I am not in that mode. I am in the mode of being deeply concerned."

But my all-time champ is Elliot Richardson, who held four major cabinet positions in the 1970s. It's hard to know where to begin picking from his trove of equivocal statements, but consider this one: "And yet, on balance, affirmative action has, I think, been a qualified success." A 13-word sentence with five hedging words. I give it first prize as the most wishy-washy sentence in modern public discourse, though a rival would be his analysis of how to ease boredom among assembly-line workers: "And so, at last, I come to the one firm conviction that I mentioned at the beginning: it is that the subject is too new for final judgments."

That's a firm conviction? Leaders who bob and weave like aging boxers don't inspire confidence—or deserve it. The same thing is true of writers. Sell yourself, and your subject will exert its own appeal. Believe in your own identity and your own opinions. Writing is an act of ego, and you might as well admit it. Use its energy to keep yourself going.

<u>5</u>

The Audience

Soon after you confront the matter of preserving your identity, another question will occur to you: "Who am I writing for?"

It's a fundamental question, and it has a fundamental answer: You are writing for yourself. Don't try to visualize the great mass audience. There is no such audience—every reader is a different person. Don't try to guess what sort of thing editors want to publish or what you think the country is in a mood to read. Editors and readers don't know what they want to read until they read it. Besides, they're always looking for something new.

Don't worry about whether the reader will "get it" if you indulge a sudden impulse for humor. If it amuses you in the act of writing, put it in. (It can always be taken out, but only you can put it in.) You are writing primarily to please yourself, and if you go about it with enjoyment you will also entertain the readers who are worth writing for. If you lose the dullards back in the dust, you don't want them anyway.

This may seem to be a paradox. Earlier I warned that the reader is an impatient bird, perched on the thin edge of distraction or sleep. Now I'm saying you must write for yourself and not be gnawed by worry over whether the reader is tagging along.

I'm talking about two different issues. One is craft, the other is attitude. The first is a question of mastering a precise skill. The second is a question of how you use that skill to express your personality.

In terms of craft, there's no excuse for losing readers through sloppy workmanship. If they doze off in the middle of your article because you have been careless about a technical detail, the fault is yours. But on the larger issue of whether the reader likes you, or likes what you are saying or how you are saying it, or agrees with it, or feels an affinity for your sense of humor or your vision of life, don't give him a moment's worry. You are who you are, he is who he is, and either you'll get along or you won't.

Perhaps this still seems like a paradox. How can you think carefully about not losing the reader and still be carefree about his opinion? I assure you that they are separate processes.

First, work hard to master the tools. Simplify, prune and strive for order. Think of this as a mechanical act, and soon your sentences will become cleaner. The act will never become as mechanical as, say, shaving or shampooing; you will always have to think about the various ways in which the tools can be used. But at least your sentences will be grounded in solid principles, and your chances of losing the reader will be smaller.

Think of the other as a creative act: the expressing of who you are. Relax and say what you want to say. And since style is who you are, you only need to be true to yourself to find it gradually emerging from under the accumulated clutter and debris, growing more distinctive every day. Perhaps the style won't solidify for years as *your* style, *your* voice. Just as it takes time to find yourself as a person, it takes time to find yourself as a stylist, and even then your style will change as you grow older.

But whatever your age, be yourself when you write. Many old men still write with the zest they had in their twenties or thirties; obviously their ideas are still young. Other old writers ramble and repeat themselves; their style is the tip-off that they have turned into garrulous bores. Many college students write as if they were desiccated alumni 30 years out. Never say anything in writing that you wouldn't comfortably say in conversation. If you're not a person who says "indeed" or "moreover," or who calls someone an individual ("he's a fine individual"), *please* don't write it.

Let's look at a few writers to see the pleasure with which they put on paper their passions and their crotchets, not caring whether the reader shares them or not. The first excerpt is from "The Hen (An Appreciation)," written by E. B. White in 1944, at the height of World War II:

Chickens do not always enjoy an honorable position among city-bred people, although the egg, I notice, goes on and on. Right now the hen is in favor. The war has deified her and she is the darling of the home front, feted at conference tables, praised in every smoking car, her girlish ways and curious habits the topic of many an excited husbandryman to whom yesterday she was a stranger without honor or allure.

My own attachment to the hen dates from 1907, and I have been faithful to her in good times and bad. Ours has not always been an easy relationship to maintain. At first, as a boy in a carefully zoned suburb, I had neighbors and police to reckon with; my chickens had to be as closely guarded as an underground newspaper. Later, as a man in the country, I had my old friends in town to reckon with, most of whom regarded the hen as a comic prop straight out of vaudeville.... Their scorn only increased my devotion to the hen. I remained loyal, as a man would to a bride whom his family received with open ridicule. Now it is my turn to wear the smile, as I listen to the enthusiastic cackling of urbanites, who have suddenly taken up the hen socially and who fill the air with their newfound ecstasy and knowledge and the relative charms of the New Hampshire Red and the Laced Wyandotte. You would think, from their nervous cries of wonder and praise, that the hen was hatched yesterday in the suburbs of New York, instead of in the remote past in the jungles of India.

To a man who keeps hens, all poultry lore is exciting and endlessly fascinating. Every spring I settle down with my farm journal and read, with the same glazed expression on my face, the age-old story of how to prepare a brooder house....

There's a man writing about a subject I have absolutely no interest in. Yet I enjoy this piece thoroughly. I like the simple beauty of its style. I like the rhythms, the unexpected but refreshing words ("deified," "allure," "cackling"), the specific details like the Laced Wyandotte and the brooder house. But mainly what I like is that this is a man telling me unabashedly about a love affair with poultry that goes back to 1907. It's written with humanity and warmth, and after three paragraphs I know quite a lot about what sort of man this hen-lover is.

Or take a writer who is almost White's opposite in terms of style, who relishes the opulent word for its opulence and doesn't deify the simple sentence. Yet they are brothers in holding firm opinions and saying what they think. This is H. L. Mencken reporting on the notorious "Monkey Trial"—the trial of John Scopes, a young teacher who taught the theory of evolution in his Tennessee classroom—in the summer of 1925:

It was hot weather when they tried the infidel Scopes at Dayton, Tenn., but I went down there very willingly, for I was eager to see something of evangelical Christianity as a going concern. In the big cities of the Republic, despite the endless efforts of consecrated men, it is laid up with a wasting disease. The very Sunday-school superintendents, taking jazz from the stealthy radio, shake their fire-proof legs; their pupils, moving into adolescence, no longer respond to the proliferating hormones by enlisting for missionary service in Africa, but resort to necking instead. Even in Dayton, I found, though the mob was up to do execution on Scopes, there was a strong smell of antinomianism. The nine churches of the village were all half empty on Sunday, and weeds choked their yards. Only two or three of the resident pastors managed to sustain themselves by their ghostly science; the rest had to take orders for mail-order pantaloons or work in the adjacent strawberry fields; one, I heard, was a barber.... Exactly twelve minutes after I reached the village I was taken in tow by a Christian man and introduced to the favorite tipple of the Cumberland Range; half corn liquor and half Coca-Cola. It seemed a dreadful dose to me, but I found that the Dayton illuminati got it down with gusto, rubbing their tummies and rolling their eyes. They were all hot for Genesis, but their faces were too florid to belong to teetotalers, and when a pretty girl came tripping down the main street, they reached for the places where their neckties should have been with all the amorous enterprise of movie stars....

This is pure Mencken in its surging momentum and its irreverence. At almost any page where you open his books he is saying something sure to outrage the professed pieties of his countrymen. The sanctity in which Americans bathed their heroes, their churches and their edifying laws—especially Prohibition—was a well of hypocrisy for him that never dried up. Some of his heaviest ammunition he hurled at politicians and Presidents—his portrait of "The Archangel Woodrow" still scorches the pages—and as for Christian believers and clerical folk, they turn up unfailingly as mountebanks and boobs.

It may seem a miracle that Mencken could get away with such heresies in the 1920s, when hero worship was an American religion and the self-righteous wrath of the Bible Belt oozed from coast to coast. Not only did he get away with it; he was the most revered and influential journalist of his generation. The impact he made on subsequent writers of nonfiction is beyond measuring, and even now his topical pieces seem as fresh as if they were written yesterday.

The secret of his popularity—aside from his pyrotechnical use of the American language—was that he was writing for himself and didn't give a damn what the reader might think. It wasn't necessary to share his prejudices to enjoy seeing them expressed with such mirthful abandon. Mencken was never timid or evasive; he didn't kowtow to the reader or curry anyone's favor. It takes courage to be such a writer, but it is out of such courage that revered and influential journalists are born.

Moving forward to our own time, here's an excerpt from *How to Survive in Your Native Land*, a book by James Herndon describing his experiences as a teacher in a California junior high school. Of all the earnest books on education that have sprouted in America, Herndon's is—for me—the one that best captures how it really is in the classroom. His style is not quite like anybody else's, but his voice is true. Here's how the book starts:

I might as well begin with Piston. Piston was, as a matter of description, a redheaded medium-sized chubby eighth-grader; his definitive characteristic was, however, stubbornness. Without going into a lot of detail, it became clear right away that what Piston didn't want to do, Piston didn't do; what Piston wanted to do, Piston did.

It really wasn't much of a problem. Piston wanted mainly to paint, draw monsters, scratch designs on mimeograph blanks and print them up, write an occasional horror story—some kids referred to him as The Ghoul—and when he didn't want to do any of those, he wanted to roam the halls and on occasion (we heard) investigate the girls' bathrooms.

We had minor confrontations. Once I wanted everyone to sit down and listen to what I had to say—something about the way they had been acting in the halls. I was letting them come and go freely and it was up to them (I planned to point out) not to raise hell so that I had to hear about it from other teachers. Sitting down was the issue—I was determined everyone was going to do it first, then I'd talk. Piston remained standing. I reordered. He paid no attention. I pointed out that I was talking to him. He indicated he heard me. I inquired then why in hell didn't he sit down. He said he didn't want to. I said I did want him to. He said that didn't matter to him. I said do it anyway. He said why? I said because I said so. He said he wouldn't. I said Look I want you to sit down and listen to what I'm going to say. He said he *was* listening. I'll listen but I won't sit down.

Well, that's the way it goes sometimes in schools. You as teacher become obsessed with an issue—I was the injured party, conferring, as usual, unheard-of freedoms, and here they were as usual taking advantage. It ain't pleasant coming in the teachers' room for coffee and having to hear somebody say that so-and-so and so-and-so from *your* class were out in the halls *without a pass* and *making faces* and *giving the finger* to kids in *my* class during the most *important* part of *my* lesson about *Egypt*—and you ought to be allowed your tendentious speech, and most everyone will allow it, sit down for it, but occasionally someone wises you up by refusing to submit where it isn't necessary.... How did any of us get into this? we ought to be asking ourselves.

Any writer who uses "ain't" and "tendentious" in the same sentence, who quotes without using quotation marks, knows what he's doing. This seemingly artless style, so full of art, is ideal for Herndon's purpose. It avoids the pretentiousness that infects so much writing by people doing worthy work, and it allows for a rich vein of humor and common sense. Herndon sounds like a good teacher and a man whose company I would enjoy. But ultimately he is writing for himself: an audience of one.

"Who am I writing for?" The question that begins this chapter has irked some readers. They want me to say "Whom am I writing for?" But I can't bring myself to say it. It's just not me.

<u>**6**</u>

<u>Words</u>

There is a kind of writing that might be called journalese, and it's the death of freshness in anybody's style. It's the common currency of newspapers and of magazines like *People*—a mixture of cheap words, made-up words and clichés that have become so pervasive that a writer can hardly help using them. You must fight these phrases or you'll sound like every hack. You'll never make your mark as a writer unless you develop a respect for words and a curiosity about their shades of meaning that is almost obsessive. The English language is rich in strong and supple words. Take the time to root around and find the ones you want.

What is "journalese"? It's a quilt of instant words patched together out of other parts of speech. Adjectives are used as nouns ("greats," "notables"). Nouns are used as verbs ("to host"), or they are chopped off to form verbs ("enthuse," "emote"), or they are padded to form verbs ("beef up," "put teeth into"). This is a world where eminent people are "famed" and their associates are "staffers," where the future is always "upcoming" and someone is forever "firing off" a note. Nobody in America has sent a note or a memo or a telegram in years. Famed diplomat Condoleezza Rice, who hosts foreign notables to beef up the morale of top State Department staffers, sits down and fires off a lot of notes. Notes that are fired off are always fired in anger and from a sitting position. What the weapon is I've never found out.

Here's an article from a famed newsmagazine that is hard to match for fatigue:

Last February, Plainclothes Patrolman Frank Serpico knocked at the door of a suspected Brooklyn heroin pusher. When the door opened a crack, Serpico shouldered his way in only to be met by a .22-cal. pistol slug crashing into his face. Somehow he survived, although there are still buzzing fragments in his head, causing dizziness and permanent deafness in his left ear. Almost as painful is the suspicion that he may well have been set up for the shooting by other policemen. For Serpico, 35, has been waging a lonely, four-year war against the routine and endemic corruption that he and others claim is rife in the New York City police department. His efforts are now sending shock waves through the ranks of New York's finest.... Though the impact of the commission's upcoming report has yet to be felt, Serpico has little hope that …

The upcoming report has yet to be felt because it's still upcoming, and as for the permanent deafness, it's a little early to tell. And what makes those buzzing fragments buzz? By now only Serpico's head should be buzzing. But apart from these lazinesses of logic, what makes the story so tired is the failure of the writer to reach for anything but the nearest cliché. "Shouldered his way," "only to be met," "crashing into his face," "waging a lonely war," "corruption that is rife," "sending shock waves," "New York's finest"—these dreary phrases constitute writing at its most banal. We know just what to expect. No surprise awaits us in the form of an unusual word, an oblique look. We are in the hands of a hack, and we know it right away. We stop reading.

Don't let yourself get in this position. The only way to avoid it is to care deeply about words. If you find yourself writing that someone recently enjoyed a spell of illness, or that a business has been enjoying a slump, ask yourself how much they enjoyed it. Notice the decisions that other writers make in their choice of words and be finicky about the ones you select from the vast supply. The race in writing is not to the swift but to the original.

Make a habit of reading what is being written today and what was written by earlier masters. Writing is learned by imitation. If anyone asked me how I learned to write, I'd say I learned by reading the men and women who were doing the kind of writing *I* wanted to do and trying to figure out how they did it. But cultivate the best models. Don't assume that because an article is in a newspaper or a magazine it must be good. Sloppy editing is common in newspapers, often for lack of time, and writers who use clichés often work for editors who have seen so many clichés that they no longer even recognize them.

Also get in the habit of using dictionaries. My favorite for handy use is *Webster's New World Dictionary*, Second College Edition, although, like all word freaks, I own bigger dictionaries that will reward me when I'm on some more specialized search. If you have any doubt of what a word means, look it up. Learn its etymology and notice what curious branches its original root has put forth. See if it has any meanings you didn't know it had. Master the small gradations between words that seem to be synonyms. What's the difference between "cajole," "wheedle," "blandish" and "coax"? Get yourself a dictionary of synonyms.

And don't scorn that bulging grab bag *Roget's Thesaurus*. It's easy to regard the book as hilarious. Look up "villain," for instance, and you'll be awash in such rascality as only a lexicographer could conjure back from centuries of iniquity, obliquity, depravity, knavery, profligacy, frailty, flagrancy, infamy, immorality, corruption, wickedness, wrongdoing, backsliding and sin. You'll find ruffians and riffraff, miscreants and malefactors, reprobates and rapscallions, hooligans and hoodlums, scamps and scapegraces, scoundrels and scalawags, jezebels and jades. You'll find adjectives to fit them all (foul and fiendish, devilish and diabolical), and adverbs and verbs to describe how the wrongdoers do their wrong, and cross-references leading to still other thickets of venality and vice. Still, there's no better friend to have around to nudge the memory than *Roget*. It saves you the time of rummaging in your brain—that network of overloaded grooves—to find the word that's right on the tip of your tongue, where it doesn't do you any good. The *Thesaurus* is to the writer what a rhyming dictionary is to the songwriter—a reminder of all the choices—and you should use it with gratitude. If, having found the scalawag and the scapegrace, you want to know how they differ, *then* go to the dictionary.

Also bear in mind, when you're choosing words and stringing them together, how they sound. This may seem absurd: readers read with their eyes. But in fact they hear what they are reading far more than you realize. Therefore such matters as rhythm and alliteration are vital to every sentence. A typical example —maybe not the best, but undeniably the nearest—is the preceding paragraph. Obviously I enjoyed making a certain arrangement of my ruffians and riffraff, my hooligans and hoodlums, and my readers enjoyed it too—far more than if I had provided a mere list. They enjoyed not only the arrangement but the effort to entertain them. They weren't enjoying it, however, with their eyes. They were hearing the words in their inner ear.

E. B. White makes the case cogently in *The Elements of Style*, a book every writer should read once a year, when he suggests trying to rearrange any phrase that has survived for a century or two, such as Thomas Paine's "These are the times that try men's souls":

Times like these try men's souls.
How trying it is to live in these times!
These are trying times for men's souls.
Soulwise, these are trying times.

Paine's phrase is like poetry and the other four are like oatmeal—which is the divine mystery of the creative process. Good writers of prose must be part poet, always listening to what they write. E. B. White is one of my favorite stylists because I'm conscious of being with a man who cares about the cadences and sonorities of the language. I relish (in my ear) the pattern his words make as they fall into a sentence. I try to surmise how in rewriting the sentence he reassembled it to end with a phrase that will momentarily linger, or how he chose one word over another because he was after a certain emotional weight. It's the difference between, say, "serene" and "tranquil"—one so soft, the other strangely disturbing because of the unusual *n* and *q*.

Such considerations of sound and rhythm should go into everything you write. If all your sentences move at the same plodding gait, which even you recognize as deadly but don't know how to cure, read them aloud. (I write entirely by ear and read everything aloud before letting it go out into the world.) You'll begin to hear where the trouble lies. See if you can gain variety by reversing the order of a sentence, or by substituting a word that has freshness or oddity, or by altering the length of your sentences so they don't all sound as if they came out of the same machine. An occasional short sentence can carry a tremendous punch. It stays in the reader's ear.

Remember that words are the only tools you've got. Learn to use them with originality and care. And also remember: somebody out there is listening.

7

Usage

All this talk about good words and bad words brings us to a gray but important area called "usage." What is good usage? What is good English? What newly minted words is it O.K. to use, and who is to be the judge? Is it O.K. to use "O.K."?

Earlier I mentioned an incident of college students hassling the administration, and in the last chapter I described myself as a word freak. Here are two fairly recent arrivals. "Hassle" is both a verb and a noun, meaning to give somebody a hard time, or the act of being given a hard time, and anyone who has ever been hassled for not properly filling out Form 35-BX will agree that the word sounds exactly right. "Freak" means an enthusiast, and there's no missing the aura of obsession that goes with calling someone a jazz freak, or a chess freak, or a sun freak, though it would probably be pushing my luck to describe a man who compulsively visits circus sideshows as a freak freak.

Anyway, I accept these two usages gladly. I don't consider them slang, or put quotation marks around them to show that I'm mucking about in the argot of the youth culture and really know better. They're good words and we need them But I won't accept "notables" and "greats" and "upcoming" and many other newcomers. They are cheap words and we *don't* need them.

Why is one word good and another word cheap? I can't give you an answer, because usage has no fixed boundaries. Language is a fabric that changes from one week to another, adding new strands and dropping old ones, and even word freaks fight over what is allowable, often reaching their decision on a wholly subjective basis such as taste ("notables" is sleazy). Which still leaves the question of who our tastemakers are.

The question was confronted in the 1960s by the editors of a brand-new dictionary, *The American Heritage Dictionary*. They assembled a "Usage Panel" to help them appraise the new words and dubious constructions that had come knocking at the door. Which ones should be ushered in, which thrown out on their ear? The panel consisted of 104 men and women—mostly writers, poets, editors and teachers—who were known for caring about the language and trying to use it well. I was a member of the panel, and over the next few years I kept getting questionnaires. Would I accept "finalize" and "escalate"? How did I feel about "It's me"? Would I allow "like" to be used as a conjunction—like so many people do? How about "mighty," as in "mighty fine"?

We were told that in the dictionary our opinions would be tabulated in a separate "Usage Note," so that readers could see how we voted. The questionnaire also left room for any comments we might feel impelled to make —an opportunity the panelists seized avidly, as we found when the dictionary was published and our comments were released to the press. Passions ran high. "Good God, no! Never!" cried Barbara W. Tuchman, asked about the verb "to author." Scholarship hath no fury like that of a language purist faced with sludge, and I shared Tuchman's vow that "author" should never be authorized, just as I agreed with Lewis Mumford that the adverb "good" should be "left as the exclusive property of Ernest Hemingway."

But guardians of usage are doing only half their job if they merely keep the language from becoming sloppy. Any dolt can rule that the suffix "wise," as in "healthwise," is doltwise, or that being "rather unique" is no more possible than being rather pregnant. The other half of the job is to help the language grow by welcoming any immigrant that will bring strength or color. Therefore I was glad that 97 percent of us voted to admit "dropout," which is clean and vivid, but that only 47 percent would accept "senior citizen," which is typical of the pudgy new intruders from the land of sociology, where an illegal alien is now an undocumented resident. I'm glad we accepted "escalate," the kind of verbal contraption I generally dislike but which the Vietnam war endowed with a precise meaning, complete with overtones of blunder.

I'm glad we took into full membership all sorts of robust words that previous dictionaries derided as "colloquial": adjectives like "rambunctious," verbs like "trigger" and "rile," nouns like "shambles" and "tycoon" and "trek," the latter approved by 78 percent to mean any difficult trip, as in "the commuter's daily trek to Manhattan." Originally it was a Cape Dutch word applied to the Boers' arduous journey by ox wagon. But our panel evidently felt that the Manhattan commuter's daily trek is no less arduous.

Still, 22 percent were unwilling to let "trek" slip into general usage. That was the virtue of revealing how our panel voted—it put our opinions on display, and writers in doubt can conduct themselves accordingly. Thus our 95 percent vote against "myself," as in "He invited Mary and myself to dinner," a word condemned as "prissy," "horrible" and "a genteelism," ought to warn off anyone who doesn't want to be prissy, horrible or genteel. As Red Smith put it, "'Myself' is the refuge of idiots taught early that 'me' is a dirty word."

On the other hand, only 66 percent of our panel rejected the verb "to contact," once regarded as tacky, and only half opposed the split infinitive and the verbs "to fault" and "to bus." So only 50 percent of your readers will fault you if you decide to voluntarily call your school board and to bus your children to another town. If you contact your school board you risk your reputation by another 16 percent. Our apparent rule of thumb was stated by Theodore M. Bernstein, author of the excellent *The Careful Writer:* "We should apply the test of convenience. Does the word fill a real need? If it does, let's give it a franchise."

All of this confirms what lexicographers have always known: that the laws of usage are relative, bending with the taste of the lawmaker. One of our panelists, Katherine Anne Porter, called "O.K." a "detestable vulgarity" and claimed she had never spoken the word in her life, whereas I freely admit that I have spoken the word "O.K." "Most," as in "most everyone," was scorned as "cute farmer talk" by Isaac Asimov and embraced as a "good English idiom" by Virgil Thomson. "Regime," meaning any administration, as in "the Truman regime," drew the approval of most everyone on the panel, as did "dynasty." But they drew the wrath of Jacques Barzun, who said, "These are technical terms, you blasted non-historians!" Probably I gave my O.K. to "regime." Now, chided by Barzun for imprecision, I think it looks like journalese. One of the words *I* railed against was "personality," as in a "TV personality." But now I wonder if it isn't the only word for that vast swarm of people who are famous for being famous— and possibly nothing else. What did the Gabor sisters actually *do?*

In the end it comes down to what is "correct" usage. We have no king to establish the King's English; we only have the President's English, which we don't want. *Webster,* long a defender of the faith, muddied the waters in 1961 with its permissive Third Edition, which argued that almost anything goes as long as somebody uses it, noting that "ain't" is "used orally in most parts of the U.S. by many cultivated speakers."

Just where *Webster* cultivated those speakers I ain't sure. Nevertheless it's true that the spoken language is looser than the written language, and *The American Heritage Dictionary* properly put its question to us in both forms. Often we allowed an oral idiom that we forbade in print as too informal, fully realizing, however, that "the pen must at length comply with the tongue," as Samuel Johnson said, and that today's spoken garbage may be tomorrow's written gold. The growing acceptance of the split infinitive, or of the preposition at the end of a sentence, proves that formal syntax can't hold the fort forever against a speaker's more comfortable way of getting the same thing said—and it shouldn't. I think a sentence is a fine thing to put a preposition at the end of.

Our panel recognized that correctness can even vary within a word. We voted heavily against "cohort" as a synonym for "colleague," except when the tone was jocular. Thus a professor would not be among his cohorts at a faculty meeting, but they would abound at his college reunion, wearing funny hats. We rejected "too" as a synonym for "very," as in "His health is not too good." Whose health is? But we approved it in sardonic or humorous use, as in "He was not too happy when she ignored him."

These may seem like picayune distinctions. They're not. They are signals to the reader that you are sensitive to the shadings of usage. "Too" when substituted for "very" is clutter: "He didn't feel too much like going shopping." But the wry example in the previous paragraph is worthy of Ring Lardner. It adds a tinge of sarcasm that otherwise wouldn't be there.

Luckily, a pattern emerged from the deliberations of our panel, and it offers a guideline that is still useful. We turned out to be liberal in accepting new words and phrases, but conservative in grammar.

It would be foolish to reject a word as perfect as "dropout," or to pretend that countless words and phrases are not entering the gates of correct usage every day, borne on the winds of science and technology, business and sports and social change: "outsource," "blog," "laptop," "mousepad," "geek," "boomer," "Google," "iPod," "hedge fund," "24/7," "multi-tasking," "slam dunk" and hundreds of others. Nor should we forget all the short words invented by the counterculture in the 1960s as a way of lashing back at the self-important verbiage of the Establishment: "trip," "rap," "crash," "trash," "funky," "split," "rip-off," "vibes," "downer," "bummer." If brevity is a prize, these were winners. The only trouble with accepting words that entered the language overnight is that they often leave just as abruptly. The "happenings" of the late 1960s no longer happen, "out of sight" is out of sight, and even "awesome" has begun to chill out. The writer who cares about usage must always know the quick from the dead.

As for the area where our Usage Panel was conservative, we upheld most of the classic distinctions in grammar—"can" and "may," "fewer" and "less," "eldest" and "oldest," etc.—and decried the classic errors, insisting that "flout" still doesn't mean "flaunt," no matter how many writers flaunt their ignorance by flouting the rule, and that "fortuitous" still means "accidental," "disinterested" still means "impartial," and "infer" doesn't mean "imply." Here we were motivated by our love of the language's beautiful precision. Incorrect usage will lose you the readers you would most like to win. Know the difference between a "reference" and an "allusion," between "connive" and "conspire," between "compare with" and "compare to." If you must use "comprise," use it right. It means "include"; dinner comprises meat, potatoes, salad and dessert.

"I choose always the grammatical form unless it sounds affected," Marianne Moore explained, and that's finally where our panel took its stand. We were not pedants, so hung up on correctness that we didn't want the language to keep refreshing itself with phrases like "hung up." But that didn't mean we had to accept every atrocity that comes lumbering in.

Meanwhile the battle continues. Today I still receive ballots from *The American Heritage Dictionary* soliciting my opinion on new locutions: verbs like "definitize" ("Congress definitized a proposal"), nouns like "affordables," colloquialisms like "the bottom line" and strays like "into" ("He's into backgammon and she's into jogging").

It no longer takes a panel of experts to notice that jargon is flooding our daily life and language. President Carter signed an executive order directing that federal regulations be written "simply and clearly." President Clinton's attorney general, Janet Reno, urged the nation's lawyers to replace "a lot of legalese" with "small, old words that all people understand"—words like "right" and "wrong" and "justice." Corporations have hired consultants to make their prose less opaque, and even the insurance industry is trying to rewrite its policies to tell us in less disastrous English what redress will be ours when disaster strikes. Whether these efforts will do much good I wouldn't want to bet. Still, there's comfort in the sight of so many watchdogs standing Canute-like on the beach, trying to hold back the tide. That's where all careful writers ought to be— looking at every new piece of flotsam that washes up and asking "Do we need it?"

I remember the first time somebody asked me, "How does that impact you?" I always thought "impact" was a noun, except in dentistry. Then I began to meet "de-impact," usually in connection with programs to de-impact the effects of some adversity. Nouns now turn overnight into verbs. We target goals and we access facts. Train conductors announce that the train won't platform. A sign on an airport door tells me that the door is alarmed. Companies are downsizing. It's part of an ongoing effort to grow the business. "Ongoing" is a jargon word whose main use is to raise morale. We face our daily job with more zest if the boss tells us it's an ongoing project; we give more willingly to institutions if they have targeted our funds for ongoing needs. Otherwise we might fall prey to disincentivization.

I could go on; I have enough examples to fill a book, but it's not a book I would want anyone to read. We're still left with the question: What is good usage? One helpful approach is to try to separate usage from jargon.

I would say, for example, that "prioritize" is jargon—a pompous new verb that sounds more important than "rank"—and that "bottom line" is usage, a metaphor borrowed from the world of bookkeeping that conveys an image we can picture. As every businessman knows, the bottom line is the one that matters. If someone says, "The bottom line is that we just can't work together," we know what he means. I don't much like the phrase, but the bottom line is that it's here to stay.

New usages also arrive with new political events. Just as Vietnam gave us "escalate," Watergate gave us a whole lexicon of words connoting obstruction and deceit, including "deep-six," "launder," "enemies list" and other "gate"-suffix scandals ("Irangate"). It's a fitting irony that under Richard Nixon "launder" became a dirty word. Today when we hear that someone laundered his funds to hide the origin of the money and the route it took, the word has a precise meaning. It's short, it's vivid, and we need it. I accept "launder" and "stonewall"; I don't accept "prioritize" and "disincentive."

I would suggest a similar guideline for separating good English from technical English. It's the difference between, say, "printout" and "input." A printout is a specific object that a computer emits. Before the advent of computers it wasn't needed; now it is. But it has stayed where it belongs. Not so with "input," which was coined to describe the information that's fed to a computer. Our input is sought on every subject, from diets to philosophical discourse ("I'd like your input on whether God really exists").

I don't want to give somebody my input and get his feedback, though I'd be glad to offer my ideas and hear what he thinks of them. Good usage, to me, consists of using good words if they already exist—as they almost always do—to express myself clearly and simply to someone else. You might say it's how I verbalize the interpersonal.

PART II

Methods

8

Unity

You learn to write by writing. It's a truism, but what makes it a truism is that it's true. The only way to learn to write is to force yourself to produce a certain number of words on a regular basis.

If you went to work for a newspaper that required you to write two or three articles every day, you would be a better writer after six months. You wouldn't necessarily be writing well; your style might still be full of clutter and clichés. But you would be exercising your powers of putting the English language on paper, gaining confidence and identifying the most common problems.

All writing is ultimately a question of solving a problem. It may be a problem of where to obtain the facts or how to organize the material. It may be a problem of approach or attitude, tone or style. Whatever it is, it has to be confronted and solved. Sometimes you will despair of finding the right solution—or any solution. You'll think, "If I live to be ninety I'll never get out of this mess." I've often thought it myself. But when I finally do solve the problem it's because I'm like a surgeon removing his 500th appendix; I've been there before.

Unity is the anchor of good writing. So, first, get your unities straight. Unity not only keeps the reader from straggling off in all directions; it satisfies your readers' subconscious need for order and reassures them that all is well at the helm. Therefore choose from among the many variables and stick to your choice.

One choice is unity of pronoun. Are you going to write in the first person, as a participant, or in the third person, as an observer? Or even in the second person, that darling of sportswriters hung up on Hemingway? ("You knew this had to be the most spine-tingling clash of giants you'd ever seen from a pressbox seat, and you weren't just some green kid who was still wet behind the ears.")

Unity of tense is another choice. Most people write mainly in the past tense ("I went up to Boston the other day"), but some people write agreeably in the present ("I'm sitting in the dining car of the Yankee Limited and we're pulling into Boston"). What is not agreeable is to switch back and forth. I'm not saying you can't use more than one tense; the whole purpose of tenses is to enable a writer to deal with time in its various gradations, from the past to the hypothetical future ("When I telephoned my mother from the Boston station, I realized that if I had written to tell her I would be coming she would have waited for me"). But you must choose the tense in which you are *principally* going to address the reader, no matter how many glances you may take backward or forward along the way.

Another choice is unity of mood. You might want to talk to the reader in the casual voice that *The New Yorker* has strenuously refined. Or you might want to approach the reader with a certain formality to describe a serious event or to present a set of important facts. Both tones are acceptable. In fact, *any* tone is acceptable. But don't mix two or three.

Such fatal mixtures are common in writers who haven't learned control. Travel writing is a conspicuous example. "My wife, Ann, and I had always wanted to visit Hong Kong," the writer begins, his blood astir with reminiscence, "and one day last spring we found ourselves looking at an airline poster and I said, 'Let's go!' The kids were grown up," he continues, and he proceeds to describe in genial detail how he and his wife stopped off in Hawaii and had such a comical time changing their money at the Hong Kong airport and finding their hotel. Fine. He is a real person taking us along on a real trip, and we can identify with him and Ann.

Suddenly he turns into a travel brochure. "Hong Kong affords many fascinating experiences to the curious sightseer," he writes. "One can ride the picturesque ferry from Kowloon and gawk at the myriad sampans as they scuttle across the teeming harbor, or take a day's trip to browse in the alleys of fabled Macao with its colorful history as a den of smuggling and intrigue. You will want to take the quaint funicular that climbs ..." Then we get back to him and Ann and their efforts to eat at Chinese restaurants, and again all is well. Everyone is interested in food, and we are being told about a personal adventure.

Then suddenly the writer is a guidebook: "To enter Hong Kong it is necessary to have a valid passport, but no visa is required. You should definitely be immunized against hepatitis and you would also be well advised to consult your physician with regard to a possible inoculation for typhoid. The climate in Hong Kong is seasonable except in July and August when ..." Our writer is gone, and so is Ann, and so—very soon—are we.

It's not that the scuttling sampans and the hepatitis shots shouldn't be included. What annoys us is that the writer never decided what kind of article he wanted to write or how he wanted to approach us. He comes at us in many guises, depending on what kind of material he is trying to purvey. Instead of controlling his material, his material is controlling him. That wouldn't happen if he took time to establish certain unities.

Therefore ask yourself some basic questions before you start. For example: "In what capacity am I going to address the reader?" (Reporter? Provider of information? Average man or woman?) "What pronoun and tense am I going to use?" "What style?" (Impersonal reportorial? Personal but formal? Personal and casual?) "What attitude am I going to take toward the material?" (Involved? Detached? Judgmental? Ironic? Amused?) "How much do I want to cover?" "What one point do I want to make?"

The last two questions are especially important. Most nonfiction writers have a definitiveness complex. They feel that they are under some obligation—to the subject, to their honor, to the gods of writing—to make their article the last word. It's a commendable impulse, but there is no last word. What you think is definitive today will turn undefinitive by tonight, and writers who doggedly pursue every last fact will find themselves pursuing the rainbow and never settling down to write. Nobody can write a book or an article "about" something. Tolstoy couldn't write a book about war and peace, or Melville a book about whaling. They made certain reductive decisions about time and place and about individual characters in that time and place—one man pursuing one whale. Every writing project must be reduced before you start to write.

Therefore think small. Decide what corner of your subject you're going to bite off, and be content to cover it well and stop. This is also a matter of energy and morale. An unwieldy writing task is a drain on your enthusiasm. Enthusiasm is the force that keeps you going and keeps the reader in your grip. When your zest begins to ebb, the reader is the first person to know it.

As for what point you want to make, every successful piece of nonfiction should leave the reader with one provocative thought that he or she didn't have before. Not two thoughts, or five—just one. So decide what single point you want to leave in the reader's mind. It will not only give you a better idea of what route you should follow and what destination you hope to reach; it will affect your decision about tone and attitude. Some points are best made by earnestness, some by dry understatement, some by humor.

Once you have your unities decided, there's no material you can't work into your frame. If the tourist in Hong Kong had chosen to write solely in the conversational vein about what he and Ann did, he would have found a natural way to weave into his narrative whatever he wanted to tell us about the Kowloon ferry and the local weather. His personality and purpose would have been intact, and his article would have held together.

Now it often happens that you'll make these prior decisions and then discover that they weren't the right ones. The material begins to lead you in an unexpected direction, where you are more comfortable writing in a different tone. That's normal—the act of writing generates some cluster of thoughts or memories that you didn't anticipate. Don't fight such a current if it feels right. Trust your material if it's taking you into terrain you didn't intend to enter but where the vibrations are good. Adjust your style accordingly and proceed to whatever destination you reach. Don't become the prisoner of a preconceived plan. Writing is no respecter of blueprints.

If this happens, the second part of your article will be badly out of joint with the first. But at least you know which part is truest to your instincts. Then it's just a matter of making repairs. Go back to the beginning and rewrite it so that your mood and your style are consistent from start to finish.

There's nothing in such a method to be ashamed of. Scissors and paste—or their equivalent on a computer—are honorable writers' tools. Just remember that all the unities must be fitted into the edifice you finally put together, however backwardly they may be assembled, or it will soon come tumbling down.

9

The Lead and the Ending

The most important sentence in any article is the first one. If it doesn't induce the reader to proceed to the second sentence, your article is dead. And if the second sentence doesn't induce him to continue to the third sentence, it's equally dead. Of such a progression of sentences, each tugging the reader forward until he is hooked, a writer constructs that fateful unit, the "lead."

How long should the lead be? One or two paragraphs? Four or five? There's no pat answer. Some leads hook the reader with just a few well-baited sentences; others amble on for several pages, exerting a slow but steady pull. Every article poses a different problem, and the only valid test is: does it work? Your lead may not be the best of all possible leads, but if it does the job it's supposed to do, be thankful and proceed.

Sometimes the length may depend on the audience you're writing for. Readers of a literary review expect its writers to start somewhat discursively, and they will stick with those writers for the pleasure of wondering where they will emerge as they move in leisurely circles toward the eventual point. But I urge you not to count on the reader to stick around. Readers want to know—very soon—what's in it for them.

Therefore your lead must capture the reader immediately and force him to keep reading. It must cajole him with freshness, or novelty, or paradox, or humor, or surprise, or with an unusual idea, or an interesting fact, or a question. Anything will do, as long as it nudges his curiosity and tugs at his sleeve.

Next the lead must do some real work. It must provide hard details that tell the reader why the piece was written and why he ought to read it. But don't dwell on the reason. Coax the reader a little more; keep him inquisitive.

Continue to build. Every paragraph should amplify the one that preceded it. Give more thought to adding solid detail and less to entertaining the reader. But take special care with the last sentence of each paragraph—it's the crucial springboard to the next paragraph. Try to give that sentence an extra twist of humor or surprise, like the periodic "snapper" in the routine of a stand-up comic. Make the reader smile and you've got him for at least one more paragraph.

Let's look at a few leads that vary in pace but are alike in maintaining pressure. I'll start with two columns of my own that first appeared in *Life* and *Look*—magazines which, judging by the comments of readers, found their consumers mainly in barbershops, hairdressing salons, airplanes and doctors' offices ("I was getting a haircut the other day and I saw your article"). I mention this as a reminder that far more periodical reading is done under the dryer than under the reading lamp, so there isn't much time for the writer to fool around.

The first is the lead of a piece called "Block That Chickenfurter":

I've often wondered what goes into a hot dog. Now I know and I wish I didn't.

Two very short sentences. But it would be hard not to continue to the second paragraph:

My trouble began when the Department of Agriculture published the hot dog's ingredients—everything that may legally qualify—because it was asked by the poultry industry to relax the conditions under which the ingredients might also include chicken. In other words, can a chickenfurter find happiness in the land of the frank?

One sentence that explains the incident that the column is based on. Then a snapper to restore the easygoing tone.

Judging by the 1,066 mainly hostile answers that the Department got when it sent out a questionnaire on this point, the very thought is unthinkable. The public mood was most felicitously caught by the woman who replied: "I don't eat feather meat of no kind."

Another fact and another smile. Whenever you're lucky enough to get a quotation as funny as that one, find a way to use it. The article then specifies what the Department of Agriculture says may go into a hot dog—a list that includes "the edible part of the muscle of cattle, sheep, swine or goats, in the diaphragm, in the heart or in the esophagus ... [but not including] the muscle found in the lips, snout or ears."

From there it progresses—not without an involuntary reflex around the esophagus—into an account of the controversy between the poultry interests and the frankfurter interests, which in turn leads to the point that Americans will eat anything that even remotely resembles a hot dog. Implicit at the end is the larger point that Americans don't know, or care, what goes into the food they eat. The style of the article has remained casual and touched with humor. But its content turns out to be more serious than readers expected when they were drawn into it by a whimsical lead.

A slower lead, luring the reader more with curiosity than with humor, introduced a piece called "Thank God for Nuts":

By any reasonable standard, nobody would want to look twice—or even once—at the piece of slippery elm bark from Clear Lake, Wisc., birthplace of pitcher Burleigh Grimes, that is on display at the National Baseball Museum and Hall of Fame in Cooperstown, N.Y. As the label explains, it is the kind of bark Grimes chewed during games "to increase saliva for throwing the spitball. When wet, the ball sailed to the plate in deceptive fashion." This would seem to be one of the least interesting facts available in America today.

But baseball fans can't be judged by any reasonable standard. We are obsessed by the minutiae of the game and nagged for the rest of our lives by the memory of players we once saw play. No item is therefore too trivial that puts us back in touch with them. I am just old enough to remember Burleigh Grimes and his well-moistened pitches sailing deceptively to the plate, and when I found his bark I studied it as intently as if I had come upon the Rosetta Stone. "So *that's* how he did it," I thought, peering at the odd botanical relic. "Slippery elm! I'll be damned."

This was only one of several hundred encounters I had with my own boyhood as I prowled through the Museum. Probably no other museum is so personal a pilgrimage to our past....

The reader is now safely hooked, and the hardest part of the writer's job is over.

One reason for citing this lead is to note that salvation often lies not in the writer's style but in some odd fact he or she was able to discover. I went up to Cooperstown and spent a whole afternoon in the museum, taking notes. Jostled everywhere by nostalgia, I gazed with reverence at Lou Gehrig's locker and Bobby Thomson's game-winning bat. I sat in a grandstand seat brought from the Polo Grounds, dug my unspiked soles into the home plate from Ebbets Field, and dutifully copied all the labels and captions that might be useful.

"These are the shoes that touched home plate as Ted finished his journey around the bases," said a label identifying the shoes worn by Ted Williams when he famously hit a home run on his last time at bat. The shoes were in much better shape than the pair—rotted open at the sides—that belonged to Walter Johnson. But the caption provided exactly the kind of justifying fact a baseball nut would want. "My feet must be comfortable when I'm out there a-pitching," the great Walter said.

The museum closed at five and I returned to my motel secure in my memories and my research. But instinct told me to go back the next morning for one more tour, and it was only then that I noticed Burleigh Grimes's slippery elm bark, which struck me as an ideal lead. It still does.

One moral of this story is that you should always collect more material than you will use. Every article is strong in proportion to the surplus of details from which you can choose the few that will serve you best—if you don't go on gathering facts forever. At some point you must stop researching and start writing.

Another moral is to look for your material everywhere, not just by reading the obvious sources and interviewing the obvious people. Look at signs and at billboards and at all the junk written along the American roadside. Read the labels on our packages and the instructions on our toys, the claims on our medicines and the graffiti on our walls. Read the fillers, so rich in self-esteem, that come spilling out of your monthly statement from the electric company and the telephone company and the bank. Read menus and catalogues and second-class mail. Nose about in obscure crannies of the newspaper, like the Sunday real estate section—you can tell the temper of a society by what patio accessories it wants. Our daily landscape is thick with absurd messages and portents. Notice them. They not only have social significance; they are often just quirky enough to make a lead that's different from everybody else's.

Speaking of everybody else's lead, there are many categories I'd be glad never to see again. One is the future archaeologist: "When some future archaeologist stumbles on the remains of our civilization, what will he make of the jukebox?" I'm tired of him already and he's not even here. I'm also tired of the visitor from Mars: "If a creature from Mars landed on our planet he would be amazed to see hordes of scantily clad earthlings lying on the sand barbecuing their skins." I'm tired of the cute event that just happened to happen "one day not long ago" or on a conveniently recent Saturday afternoon: "One day not long ago a small button-nosed boy was walking with his dog, Terry, in a field outside Paramus, N.J., when he saw something that looked strangely like a balloon rising out of the ground." And I'm very tired of the have-in-common lead: "What did Joseph Stalin, Douglas MacArthur, Ludwig Wittgenstein, Sherwood Anderson,

Jorge Luis Borges and Akira Kurosawa have in common? They all loved Westerns." Let's retire the future archaeologist and the man from Mars and the button-nosed boy. Try to give your lead a freshness of perception or detail.

Consider this lead, by Joan Didion, on a piece called "7000 Romaine, Los Angeles 38":

Seven Thousand Romaine Street is in that part of Los Angeles familiar to admirers of Raymond Chandler and Dashiell Hammett: the underside of Hollywood, south of Sunset Boulevard, a middle-class slum of "model studios" and warehouses and two-family bungalows. Because Paramount and Columbia and Desilu and the Samuel Goldwyn studios are nearby, many of the people who live around here have some tenuous connection with the motion-picture industry. They once processed fan photographs, say, or knew Jean Harlow's manicurist. 7000 Romaine looks itself like a faded movie exterior, a pastel building with chipped *art moderne* detailing, the windows now either boarded or paned with chicken-wire glass and, at the entrance, among the dusty oleander, a rubber mat that reads WELCOME.

Actually no one is welcome, for 7000 Romaine belongs to Howard Hughes, and the door is locked. That the Hughes "communications center" should lie here in the dull sunlight of Hammett-Chandler country is one of those circumstances that satisfy one's suspicion that life is indeed a scenario, for the Hughes empire has been in our time the only industrial complex in the world—involving, over the years, machinery manufacture, foreign oil-tool subsidiaries, a brewery, two airlines, immense real-estate holdings, a major motion-picture studio, and an electronics and missile operation—run by a man whose *modus operandi* most closely resembles that of a character in *The Big Sleep.*

As it happens, I live not far from 7000 Romaine, and I make a point of driving past it every now and then, I suppose in the same spirit that Arthurian scholars visit the Cornish coast. I am interested in the folklore of Howard Hughes....

What is pulling us into this article—toward, we hope, some glimpse of how Hughes operates, some hint of the riddle of the Sphinx—is the steady accumulation of facts that have pathos and faded glamour. Knowing Jean Harlow's manicurist is such a minimal link to glory, the unwelcoming welcome mat such a queer relic of a golden age when Hollywood's windows weren't paned with chicken-wire glass and the roost was ruled by giants like Mayer and DeMille and Zanuck, who could actually be seen exercising their mighty power. We want to know more; we read on.

Another approach is to just tell a story. It's such a simple solution, so obvious and unsophisticated, that we often forget that it's available to us. But narrative is the oldest and most compelling method of holding someone's attention; everybody wants to be told a story. Always look for ways to convey your information in narrative form. What follows is the lead of Edmund Wilson's account of the discovery of the Dead Sea Scrolls, one of the most astonishing relics of antiquity to turn up in modern times. Wilson doesn't spend any time setting the stage. This is not the "breakfast-to-bed" format used by inexperienced writers, in which a fishing trip begins with the ringing of an alarm clock before daylight. Wilson starts right in—whap!—and we are caught:

At some point rather early in the spring of 1947, a Bedouin boy called Muhammed the Wolf was minding some goats near a cliff on the western shore of the Dead Sea. Climbing up after one that had strayed, he noticed a cave that he had not seen before, and he idly threw a stone into it. There was an unfamiliar sound of breakage. The boy was frightened and ran away. But he later came back with another boy, and together they explored the cave. Inside were several tall clay jars, among fragments of other jars. When they took off the bowl-like lids, a very bad smell arose, which came from dark oblong lumps that were found inside all the jars. When they got these lumps out of the cave, they saw that they were wrapped up in lengths of linen and coated with a black layer of what seemed to be pitch or wax. They unrolled them and found long manuscripts, inscribed in parallel columns on thin sheets that had been sewn together. Though these manuscripts had faded and crumbled in places, they were in general remarkably clear. The character, they saw, was not Arabic. They wondered at the scrolls and kept them, carrying them along when they moved.

These Bedouin boys belonged to a party of contrabanders, who had been smuggling their goats and other goods out of Transjordan into Palestine. They had detoured so far to the south in order to circumvent the Jordan bridge, which the customs officers guarded with guns, and had floated their commodities across the stream. They were now on their way to Bethlehem to sell their stuff in the black market....

Yet there can be no firm rules for how to write a lead. Within the broad rule of not letting the reader get away, all writers must approach their subject in a manner that most naturally suits what they are writing about and who they are. Sometimes you can tell your whole story in the first sentence. Here's the opening sentence of seven memorable nonfiction books:

In the beginning God created heaven and earth.

—THE BIBLE

In the summer of the Roman year 699, now described as the year 55 before the birth of Christ, the Proconsul of Gaul, Gaius Julius Caesar, turned his gaze upon Britain.

—WINSTON S. CHURCHILL, A HISTORY OF THE ENGLISH-SPEAKING
PEOPLES

Put this puzzle together and you will find milk, cheese and eggs, meat, fish, beans and cereals, greens, fruits and root vegetables—foods that contain our essential daily needs.

—IRMA S. ROMBAUER, JOY OF COOKING

To the Manus native the world is a great platter, curving upwards on all sides, from his flat lagoon village where the pile-houses stand like long-legged birds, placid and unstirred by the changing tides.

—MARGARET MEAD, GROWING UP IN NEW GUINEA

The problem lay buried, unspoken, for many years in the minds of American women.

—BETTY FRIEDAN, THE FEMININE MYSTIQUE

Within five minutes, or ten minutes, no more than that, three of the others had called her on the telephone to ask her if she had heard that something had happened out there.

—TOM WOLFE, THE RIGHT STUFF

You know more than you think you do.

—BENJAMIN SPOCK, BABY AND CHILD CARE

Those are some suggestions on how to get started. Now I want to tell you how to stop. Knowing when to end an article is far more important than most writers realize. You should give as much thought to choosing your last sentence as you did to your first. Well, almost as much.

That may seem hard to believe. If your readers have stuck with you from the beginning, trailing you around blind corners and over bumpy terrain, surely they won't leave when the end is in sight. Surely they will, because the end that's in sight turns out to be a mirage. Like the minister's sermon that builds to a series of perfect conclusions that never conclude, an article that doesn't stop where it should stop becomes a drag and therefore a failure.

Most of us are still prisoners of the lesson pounded into us by the composition teachers of our youth: that every story must have a beginning, a middle and an end. We can still visualize the outline, with its Roman numerals (I, II and III), which staked out the road we would faithfully trudge, and its subnumerals (IIa and IIb) denoting lesser paths down which we would briefly poke. But we always promised to get back to III and summarize our journey.

That's all right for elementary and high school students uncertain of their ground. It forces them to see that every piece of writing should have a logical design. It's a lesson worth knowing at any age—even professional writers are adrift more often than they would like to admit. But if you're going to write good nonfiction you must wriggle out of III's dread grip.

You'll know you have arrived at III when you see emerging on your screen a sentence that begins, "In sum, it can be noted that …" Or a question that asks, "What insights, then, have we been able to glean from …?" These are signals that you are about to repeat in compressed form what you have already said in detail. The reader's interest begins to falter; the tension you have built begins to sag. Yet you will be true to Miss Potter, your teacher, who made you swear fealty to the holy outline. You remind the reader of what can, in sum, be noted. You go gleaning one more time in insights you have already adduced.

But your readers hear the laborious sound of cranking. They notice what you are doing and how bored you are by it. They feel the stirrings of resentment. Why didn't you give more thought to how you were going to wind this thing up? Or are you summarizing because you think they're too dumb to get the point? Still, you keep cranking. But the readers have another option. They quit.

That's the negative reason for remembering the importance of the last sentence. Failure to know where that sentence should occur can wreck an article that until its final stage has been tightly constructed. The positive reason for ending well is that a good last sentence—or last paragraph—is a joy in itself. It gives the reader a lift, and it lingers when the article is over.

The perfect ending should take your readers slightly by surprise and yet seem exactly right. They didn't expect the article to end so soon, or so abruptly, or to say what it said. But they know it when they see it. Like a good lead, it works. It's like the curtain line in a theatrical comedy. We are in the middle of a scene (we think), when suddenly one of the actors says something funny, or outrageous, or epigrammatic, and the lights go out. We are startled to find the scene over, and then delighted by the aptness of how it ended. What delights us is the playwright's perfect control.

For the nonfiction writer, the simplest way of putting this into a rule is: when you're ready to stop, stop. If you have presented all the facts and made the point you want to make, look for the nearest exit.

Often it takes just a few sentences to wrap things up. Ideally they should encapsulate the idea of the piece and conclude with a sentence that jolts us with its fitness or unexpectedness. Here's how H. L. Mencken ends his appraisal of President Calvin Coolidge, whose appeal to the "customers" was that his "government governed hardly at all; thus the ideal of Jefferson was realized at last, and the Jeffersonians were delighted":

We suffer most, not when the White House is a peaceful dormitory, but when it [has] a tin-pot Paul bawling from the roof. Counting out Harding as a cipher only, Dr. Coolidge was preceded by one World Saver and followed by two more. What enlightened American, having to choose between any of them and another Coolidge, would hesitate for an instant? There were no thrills while he reigned, but neither were there any headaches. He had no ideas, and he was not a nuisance.

The five short sentences send the reader on his way quickly and with an arresting thought to take along. The notion of Coolidge having no ideas and not being a nuisance can't help leaving a residue of enjoyment. It works.

Something I often do in my writing is to bring the story full circle—to strike at the end an echo of a note that was sounded at the beginning. It gratifies my sense of symmetry, and it also pleases the reader, completing with its resonance the journey we set out on together.

But what usually works best is a quotation. Go back through your notes to find some remark that has a sense of finality, or that's funny, or that adds an unexpected closing detail. Sometimes it will jump out at you during the interview—I've often thought, "That's my ending!"—or during the process of writing. In the mid-1960s, when Woody Allen was just becoming established as America's resident neurotic, doing nightclub monologues, I wrote the first long magazine piece that took note of his arrival. It ended like this:

"If people come away relating to me as a person," Allen says, "rather than just enjoying my jokes; if they come away wanting to hear me again, no matter what I might talk about, then I'm succeeding." Judging by the returns, he is. Woody Allen is Mr. Related-To, and he seems a good bet to hold the franchise for many years.

Yet he does have a problem all his own, unshared by, unrelated to, the rest of America. "I'm obsessed," he says, "by the fact that my mother genuinely resembles Groucho Marx."

There's a remark from so far out in left field that nobody could see it coming. The surprise it carries is tremendous. How could it not be a perfect ending? Surprise is the most refreshing element in nonfiction writing. If something surprises you it will also surprise—and delight—the people you are writing for, especially as you conclude your story and send them on their way.

10

Bits & Pieces

This is a chapter of scraps and morsels—small admonitions on many points that I have collected under one, as they say, umbrella.

VERBS.

Use active verbs unless there is no comfortable way to get around using a passive verb. The difference between an activeverb style and a passive-verb style —in clarity and vigor—is the difference between life and death for a writer.

"Joe saw him" is strong. "He was seen by Joe" is weak. The first is short and precise; it leaves no doubt about who did what. The second is necessarily longer and it has an insipid quality: something was done by somebody to someone else. It's also ambiguous. How often was he seen by Joe? Once? Every day? Once a week? A style that consists of passive constructions will sap the reader's energy. Nobody ever quite knows what is being perpetrated by whom and on whom.

I use "perpetrated" because it's the kind of word that passive-voice writers are fond of. They prefer long words of Latin origin to short Anglo-Saxon words— which compounds their trouble and makes their sentences still more glutinous. Short is better than long. Of the 701 words in Lincoln's Second Inaugural Address, a marvel of economy in itself, 505 are words of one syllable and 122 are words of two syllables.

Verbs are the most important of all your tools. They push the sentence forward and give it momentum. Active verbs push hard; passive verbs tug fitfully. Active verbs also enable us to visualize an activity because they require a pronoun ("he"), or a noun ("the boy"), or a person ("Mrs. Scott") to put them in motion. Many verbs also carry in their imagery or in their sound a suggestion of what they mean: glitter, dazzle, twirl, beguile, scatter, swagger, poke, pamper, vex. Probably no other language has such a vast supply of verbs so bright with color. Don't choose one that is dull or merely serviceable. Make active verbs activate your sentences, and avoid the kind that need an appended preposition to complete their work. Don't set up a business that you can start or launch. Don't say that the president of the company stepped down. Did he resign? Did he retire? Did he get fired? Be precise. Use precise verbs.

If you want to see how active verbs give vitality to the written word, don't just go back to Hemingway or Thurber or Thoreau. I commend the King James Bible and William Shakespeare.

ADVERBS.

Most adverbs are unnecessary. You will clutter your sentence and annoy the reader if you choose a verb that has a specific meaning and then add an adverb that carries the same meaning. Don't tell us that the radio blared loudly; "blare" connotes loudness. Don't write that someone clenched his teeth tightly; there's no other way to clench teeth. Again and again in careless writing, strong verbs are weakened by redundant adverbs. So are adjectives and other parts of speech: "effortlessly easy," "slightly spartan," "totally flabbergasted." The beauty of "flabbergasted" is that it implies an astonishment that is total; I can't picture someone being partly flabbergasted. If an action is so easy as to be effortless, use "effortless." And what is "slightly spartan"? Perhaps a monk's cell with wall-to-wall carpeting. Don't use adverbs unless they do necessary work. Spare us the news that the winning athlete grinned widely.

And while we're at it, let's retire "decidedly" and all its slippery cousins. Every day I see in the paper that some situations are decidedly better and others are decidedly worse, but I never know how decided the improvement is, or who did the deciding, just as I never know how eminent a result is that's eminently fair, or whether to believe a fact that's arguably true. "He's arguably the best pitcher on the Mets," the preening sportswriter writes, aspiring to Parnassus, which Red Smith reached by never using words like "arguably." Is the pitcher—it can be proved by argument—the best pitcher on the team? If so, please omit "arguably." Or is he *perhaps*—the opinion is open to argument—the best pitcher? Admittedly I don't know. It's virtually a toss-up.

ADJECTIVES.

Most adjectives are also unnecessary. Like adverbs, they are sprinkled into sentences by writers who don't stop to think that the concept is already in the noun. This kind of prose is littered with precipitous cliffs and lacy spiderwebs, or with adjectives denoting the color of an object whose color is well known: yellow daffodils and brownish dirt. If you want to make a value judgment about daffodils, choose an adjective like "garish." If you're in a part of the country where the dirt is red, feel free to mention the red dirt. Those adjectives would do a job that the noun alone wouldn't be doing.

Most writers sow adjectives almost unconsciously into the soil of their prose to make it more lush and pretty, and the sentences become longer and longer as they fill up with stately elms and frisky kittens and hard-bitten detectives and sleepy lagoons. This is adjective-by-habit—a habit you should get rid of. Not every oak has to be gnarled. The adjective that exists solely as decoration is a self-indulgence for the writer and a burden for the reader.

Again, the rule is simple: make your adjectives do work that needs to be done. "He looked at the gray sky and the black clouds and decided to sail back to the harbor." The darkness of the sky and the clouds is the reason for the decision. If it's important to tell the reader that a house was drab or a girl was beautiful, by all means use "drab" and "beautiful." They will have their proper power because you have learned to use adjectives sparsely.

Little Qualifiers.

Prune out the small words that qualify how you feel and how you think and what you saw: "a bit," "a little," "sort of," "kind of," "rather," "quite," "very," "too," "pretty much," "in a sense" and dozens more. They dilute your style and your persuasiveness.

Don't say you were a bit confused and sort of tired and a little depressed and somewhat annoyed. Be confused. Be tired. Be depressed. Be annoyed. Don't hedge your prose with little timidities. Good writing is lean and confident.

Don't say you weren't too happy because the hotel was pretty expensive. Say you weren't happy because the hotel was expensive. Don't tell us you were quite fortunate. How fortunate is that? Don't describe an event as rather spectacular or very awesome. Words like "spectacular" and "awesome" don't submit to measurement. "Very" is a useful word to achieve emphasis, but far more often it's clutter. There's no need to call someone very methodical. Either he is methodical or he isn't.

The large point is one of authority. Every little qualifier whittles away some fraction of the reader's trust. Readers want a writer who believes in himself and in what he is saying. Don't diminish that belief. Don't be kind of bold. Be bold.

Punctuation.

These are brief thoughts on punctuation, in no way intended as a primer. If you don't know how to punctuate—and many college students still don't—get a grammar book.

The Period. There's not much to be said about the period except that most writers don't reach it soon enough. If you find yourself hopelessly mired in a long sentence, it's probably because you're trying to make the sentence do more than it can reasonably do—perhaps express two dissimilar thoughts. The quickest way out is to break the long sentence into two short sentences, or even three. There is no minimum length for a sentence that's acceptable in the eyes of God. Among good writers it is the short sentence that predominates, and don't tell me about Norman Mailer—he's a genius. If you want to write long sentences, be a genius. Or at least make sure that the sentence is under control from beginning to end, in syntax and punctuation, so that the reader knows where he is at every step of the winding trail.

The Exclamation Point. Don't use it unless you must to achieve a certain effect. It has a gushy aura, the breathless excitement of a debutante commenting on an event that was exciting only to her: "Daddy says I must have had too much champagne!" "But honestly, I could have danced all night!" We have all suffered more than our share of these sentences in which an exclamation point knocks us over the head with how cute or wonderful something was. Instead, construct your sentence so that the order of the words will put the emphasis where you want it. Also resist using an exclamation point to notify the reader that you are making a joke or being ironic. "It never occurred to me that the water pistol might be loaded!" Readers are annoyed by your reminder that this was a comical moment. They are also robbed of the pleasure of finding it funny on their own. Humor is best achieved by understatement, and there's nothing subtle about an exclamation point.

The Semicolon. There is a 19th-century mustiness that hangs over the semicolon. We associate it with the carefully balanced sentences, the judicious weighing of "on the one hand" and "on the other hand," of Conrad and Thackeray and Hardy. Therefore it should be used sparingly by modern writers of nonfiction. Yet I notice that it turns up quite often in the passages I've quoted in this book and that I use it often myself—usually to add a related thought to the first half of a sentence. Still, the semicolon brings the reader, if not to a halt, at least to a pause. So use it with discretion, remembering that it will slow to a Victorian pace the early-21st-century momentum you're striving for, and rely instead on the period and the dash.

The Dash. Somehow this invaluable tool is widely regarded as not quite proper—a bumpkin at the genteel dinner table of good English. But it has full membership and will get you out of many tight corners. The dash is used in two ways. One is to amplify or justify in the second part of the sentence a thought you stated in the first part. "We decided to keep going—it was only 100 miles more and we could get there in time for dinner." By its very shape the dash pushes the sentence ahead and explains why they decided to keep going. The other use involves two dashes, which set apart a parenthetical thought within a longer sentence. "She told me to get in the car—she had been after me all summer to have a haircut—and we drove silently into town." An explanatory detail that might otherwise have required a separate sentence is neatly dispatched along the way.

The Colon. The colon has begun to look even more antique than the semicolon, and many of its functions have been taken over by the dash. But it still serves well its pure role of bringing your sentence to a brief halt before you plunge into, say, an itemized list. "The brochure said the ship would stop at the following ports: Oran, Algiers, Naples, Brindisi, Piraeus, Istanbul and Beirut." You can't beat the colon for work like that.

MOOD CHANGERS.

Learn to alert the reader as soon as possible to any change in mood from the previous sentence. At least a dozen words will do this job for you: "but," "yet," "however," "nevertheless," "still," "instead," "thus," "therefore," "meanwhile," "now," "later," "today," "subsequently" and several more. I can't overstate how much easier it is for readers to process a sentence if you start with "but" when you're shifting direction. Or, conversely, how much harder it is if they must wait until the end to realize that you have shifted.

Many of us were taught that no sentence should begin with "but." If that's what you learned, unlearn it—there's no stronger word at the start. It announces total contrast with what has gone before, and the reader is thereby primed for the change. If you need relief from too many sentences beginning with "but," switch to "however." It is, however, a weaker word and needs careful placement. Don't start a sentence with "however"—it hangs there like a wet dishrag. And don't end with "however"—by that time it has lost its howeverness. Put it as early as you reasonably can, as I did three sentences ago. Its abruptness then becomes a virtue.

"Yet" does almost the same job as "but," though its meaning is closer to "nevertheless." Either of those words at the beginning of a sentence—"Yet he decided to go" or "Nevertheless he decided to go"—can replace a whole long phrase that summarizes what the reader has just been told: *"Despite the fact that all these dangers had been pointed out to him,* he decided to go." Look for all the places where one of these short words will instantly convey the same meaning as a long and dismal clause. "Instead I took the train." "Still I had to admire him." "Thus I learned how to smoke." "It was therefore easy to meet him." "Meanwhile I had talked to John." What a vast amount of huffing and puffing these pivotal words save! (The exclamation point is to show that I really mean it.)

As for "meanwhile," "now," "today" and "later," what they also save is confusion, for careless writers often change their time frame without remembering to tip the reader off. "Now I know better." "Today you can't find such an item." "Later I found out why." Always make sure your readers are oriented. Always ask yourself where you left them in the previous sentence.

CONTRACTIONS.

Your style will be warmer and truer to your personality if you use contractions like "I'll" and "won't" and "can't" when they fit comfortably into what you're writing. "I'll be glad to see them if they don't get mad" is less stiff than "I will be glad to see them if they do not get mad." (Read that aloud and hear how stilted it sounds.) There's no rule against such informality—trust your ear and your instincts. I only suggest avoiding one form—"I'd," "he'd," "we'd," etc.— because "I'd" can mean both "I had" and "I would," and readers can get well into a sentence before learning which meaning it is. Often it's not the one they thought it was. Also, don't invent contractions like "could've." They cheapen your style. Stick with the ones you can find in the dictionary.

THAT AND WHICH.

Anyone who tries to explain "that" and "which" in less than an hour is asking for trouble. Fowler, in his *Modern English Usage*, takes 25 columns of type. I'm going for two minutes, perhaps the world record. Here (I hope) is much of what you need to bear in mind:

Always use "that" unless it makes your meaning ambiguous. Notice that in carefully edited magazines, such as *The New Yorker*, "that" is by far the predominant usage. I mention this because it is still widely believed—a residue from school and college—that "which" is more correct, more acceptable, more literary. It's not. In most situations, "that" is what you would naturally say and therefore what you should write.

If your sentence needs a comma to achieve its precise meaning, it probably needs "which." "Which" serves a particular identifying function, different from "that." (A) "Take the shoes that are in the closet." This means: take the shoes that are in the closet, not the ones under the bed. (B) "Take the shoes, which are in the closet." Only one pair of shoes is under discussion; the "which" usage tells you where they are. Note that the comma is necessary in B, but not in A.

A high proportion of "which" usages narrowly describe, or identify, or locate, or explain, or otherwise qualify the phrase that preceded the comma:

The house, which has a red roof,
The store, which is called Bob's Hardware,
The Rhine, which is in Germany,
The monsoon, which is a seasonal wind,
The moon, which I saw from the porch,

That's all I'm going to say that I think you initially need to know to write good nonfiction, which is a form that requires exact marshaling of information.

Concept Nouns.

Nouns that express a concept are commonly used in bad writing instead of verbs that tell what somebody did. Here are three typical dead sentences:

The common reaction is incredulous laughter.
Bemused cynicism isn't the only response to the old system.
The current campus hostility is a symptom of the change.

What is so eerie about these sentences is that they have no people in them. They also have no working verbs—only "is" or "isn't." The reader can't visualize anybody performing some activity; all the meaning lies in impersonal nouns that embody a vague concept: "reaction," "cynicism," "response," "hostility." Turn these cold sentences around. Get people doing things:

Most people just laugh with disbelief.
Some people respond to the old system by turning cynical; others say ...
It's easy to notice the change—you can see how angry all the students are.

My revised sentences aren't jumping with vigor, partly because the material I'm trying to knead into shape is shapeless dough. But at least they have real people and real verbs. Don't get caught holding a bag full of abstract nouns. You'll sink to the bottom of the lake and never be seen again.

Creeping Nounism.

This is a new American disease that strings two or three nouns together where one noun—or, better yet, one verb—will do. Nobody goes broke now; we have money problem areas. It no longer rains; we have precipitation activity or a thunderstorm probability situation. Please, let it rain.

Today as many as four or five concept nouns will attach themselves to each other, like a molecule chain. Here's a brilliant specimen I recently found: "Communication facilitation skills development intervention." Not a person in sight, or a working verb. I think it's a program to help students write better.

OVERSTATEMENT.

"The living room looked as if an atomic bomb had gone off there," writes the novice writer, describing what he saw on Sunday morning after a party that got out of hand. Well, we all know he's exaggerating to make a droll point, but we also know that an atomic bomb *didn't* go off there, or any other bomb except maybe a water bomb. "I felt as if ten 747 jets were flying through my brain," he writes, "and I seriously considered jumping out the window and killing myself." These verbal high jinks can get just so high—and this writer is already well over the limit—before the reader feels an overpowering drowsiness. It's like being trapped with a man who can't stop reciting limericks. Don't overstate. You didn't really consider jumping out the window. Life has more than enough truly horrible funny situations. Let the humor sneak up so we hardly hear it coming.

CREDIBILITY.

Credibility is just as fragile for a writer as for a President. Don't inflate an incident to make it more outlandish than it actually was. If the reader catches you in just one bogus statement that you are trying to pass off as true, everything you write thereafter will be suspect. It's too great a risk, and not worth taking.

DICTATION.

Much of the "writing" done in America is done by dictation. Administrators, executives, managers, educators and other officials think in terms of using their time efficiently. They think the quickest way of getting something "written" is to dictate it to a secretary and never look at it. This is false economy—they save a few hours and blow their whole personality. Dictated sentences tend to be pompous, sloppy and redundant. Executives who are so busy that they can't avoid dictating should at least find time to edit what they have dictated, crossing words out and putting words in, making sure that what they finally write is a true reflection of who they are, especially if it's a document that will go to customers who will judge their personality and their company on the basis of their style.

Writing Is Not a Contest.

Every writer is starting from a different point and is bound for a different destination. Yet many writers are paralyzed by the thought that they are competing with everybody else who is trying to write and presumably doing it better. This can often happen in a writing class. Inexperienced students are chilled to find themselves in the same class with students whose byline has appeared in the college newspaper. But writing for the college paper is no great credential; I've often found that the hares who write for the paper are overtaken by the tortoises who move studiously toward the goal of mastering the craft. The same fear hobbles freelance writers, who see the work of other writers appearing in magazines while their own keeps returning in the mail. Forget the competition and go at your own pace. Your only contest is with yourself.

The Subconscious Mind.

Your subconscious mind does more writing than you think. Often you'll spend a whole day trying to fight your way out of some verbal thicket in which you seem to be tangled beyond salvation. Frequently a solution will occur to you the next morning when you plunge back in. While you slept, your writer's mind didn't. A writer is always working. Stay alert to the currents around you. Much of what you see and hear will come back, having percolated for days or months or even years through your subconscious mind, just when your conscious mind, laboring to write, needs it.

The Quickest Fix.

Surprisingly often a difficult problem in a sentence can be solved by simply getting rid of it. Unfortunately, this solution is usually the last one that occurs to writers in a jam. First they will put the troublesome phrase through all kinds of exertions—moving it to some other part of the sentence, trying to rephrase it, adding new words to clarify the thought or to oil whatever is stuck. These efforts only make the situation worse, and the writer is left to conclude that there *is* no solution to the problem—not a comforting thought. When you find yourself at such an impasse, look at the troublesome element and ask, "Do I need it at all?" Probably you don't. It was trying to do an unnecessary job all along—that's why it was giving you so much grief. Remove it and watch the afflicted sentence spring to life and breathe normally. It's the quickest cure and often the best.

Paragraphs.

Keep your paragraphs short. Writing is visual—it catches the eye before it has a chance to catch the brain. Short paragraphs put air around what you write and make it look inviting, whereas a long chunk of type can discourage a reader from even starting to read.

Newspaper paragraphs should be only two or three sentences long; newspaper type is set in a narrow width, and the inches quickly add up. You may think such frequent paragraphing will damage the development of your point. Obviously *The New Yorker* is obsessed by this fear—a reader can go for miles without relief. Don't worry; the gains far outweigh the hazards.

But don't go berserk. A succession of tiny paragraphs is as annoying as a paragraph that's too long. I'm thinking of all those midget paragraphs—verbless wonders—written by modern journalists to make their articles quick 'n' easy. Actually they make the reader's job harder by chopping up a natural train of thought. Compare the following two arrangements of the same article—how they look at a glance and how they read:

<table>
<tr>
<td>

The No. 2 lawyer at the White House left work early on Tuesday, drove to an isolated park overlooking the Potomac River and took his life.

A revolver in his hand, slumped against a Civil War–era cannon, he left behind no note, no explanation.

Only friends, family and colleagues in stunned sorrow.

And a life story that until Tuesday had read like any man's fantasy.

</td>
<td>

The No. 2 lawyer at the White House left work early on Tuesday, drove to an isolated park overlooking the Potomac River and took his life. A revolver in his hand, slumped against a Civil War–era cannon, he left behind no note, no explanation—only friends, family and colleagues in stunned sorrow. He also left behind a life story that until Tuesday had read like any man's fantasy.

</td>
</tr>
</table>

The Associated Press version *(left)*, with its breezy paragraphing and verbless third and fourth sentences, is disruptive and condescending. "Yoo-hoo! Look how simple I'm making this for you!" the reporter is calling to us. My version *(right)* gives the reporter the dignity of writing good English and building three sentences into a logical unit.

Paragraphing is a subtle but important element in writing nonfiction articles and books—a road map constantly telling your reader how you have organized your ideas. Study good nonfiction writers to see how they do it. You'll find that almost all of them think in paragraph units, not in sentence units. Each paragraph has its own integrity of content and structure.

SEXISM.

One of the most vexing new questions for writers is what to do about sexist language, especially the "he-she" pronoun. The feminist movement helpfully revealed how much sexism lurks in our language, not only in the offensive "he" but in the hundreds of words that carry an invidious meaning or some overtone of judgment. They are words that patronize ("gal"), or that imply second-class status ("poetess"), or a second-class role ("housewife"), or a certain kind of empty-headedness ("the girls"), or that demean the ability of a woman to do a certain kind of job ("lady lawyer"), or that are deliberately prurient ("divorcée," "coed," "blonde") and are seldom applied to men. Men get mugged; a woman who gets mugged is a shapely stewardess or a pert brunette.

More damaging—and more subtle—are all the usages that treat women as possessions of the family male, not as people with their own identity who played an equal part in the family saga: "Early settlers pushed west with their wives and children." Turn those settlers into pioneer families, or pioneer couples who went west with their sons and daughters, or men and women who settled the West. Today there are very few roles that aren't open to both sexes. Don't use constructions that suggest that only men can be settlers or farmers or cops or firefighters.

A thornier problem is raised by the feminists' annoyance with words that contain "man," such as "chairman" and "spokesman." Their point is that women can chair a committee as well as a man and are equally good at spoking. Hence the flurry of new words like "chairperson" and "spokeswoman." Those makeshift words from the 1960s raised our consciousness about sex discrimination, both in words and in attitudes. But in the end they are makeshift words, sometimes hurting the cause more than helping it. One solution is to find another term: "chair" for "chairman," "company representative" for "spokesman." You can also convert the noun into a verb: "Speaking for the company, Ms. Jones said ..." Where a certain occupation has both a masculine and a feminine form, look for a generic substitute. Actors and actresses can become performers.

This still leaves the bothersome pronoun. "He" and "him" and "his" are words that rankle. "Every employee should decide what he thinks is best for him and his dependents." What are we to do about these countless sentences? One solution is to turn them into the plural: "All employees should decide what they think is best for them and their dependents." But this is good only in small doses.

A style that converts every "he" into a "they" will quickly turn to mush.

Another common solution is to use "or": "Every employee should decide what he or she thinks is best for him or her." But again, it should be used sparingly. Often a writer will find several situations in an article where he or she can use "he or she," or "him or her," if it seems natural. By "natural" I mean that the writer is serving notice that he (or she) has the problem in mind and is trying his (or her) best within reasonable limits. But let's face it: the English language is stuck with the generic masculine ("Man shall not live by bread alone"). To turn every "he" into a "he or she," and every "his" into a "his or her," would clog the language.

In early editions of *On Writing Well* I used "he" and "him" to refer to "the reader," "the writer," "the critic," "the humorist," *etc.* I felt that the book would be harder to read if I used "he or she" with every such mention. (I reject "he/she" altogether; the slant has no place in good English.) Over the years, however, many women wrote to nudge me about this. They said that as writers and readers themselves they resent always having to visualize a man doing the writing and reading, and they're right; I stand nudged. Most of the nudgers urged me to adopt the plural: to use "readers" and "writers," followed thereafter by "they." I don't like plurals; they weaken writing because they are less specific than the singular, less easy to visualize. I'd like every writer to visualize *one* reader struggling to read what he or she has written. Nevertheless I found three or four hundred places where I could eliminate "he," "him," "his," "himself" or "man," mainly by switching to the plural, with no harm done; the sky didn't fall in. Where the male pronoun remains in this edition I felt it was the only solution that wasn't cumbersome.

The best solutions simply eliminate "he" and its connotations of male ownership by using other pronouns or by altering some other component of the sentence. "We" is a handy replacement for "he." "Our" and "the" can often replace "his." (A) "First *he* notices what's happening to *his* kids and he blames it on *his* neighborhood." (B) "First *we* notice what's happening to *our* kids and we blame it on *the* neighborhood." General nouns can replace specific nouns. (A) "Doctors often neglect their wives and children." (B) "Doctors often neglect their families." Countless sins can be erased by such small changes.

One other pronoun that helped me in my repairs was "you." Instead of talking about what "the writer" does and the trouble *he* gets into, I found more places where I could address the writer directly ("You'll often find ..."). It doesn't work for every kind of writing, but it's a godsend to anyone writing an instructional book or a self-help book. The voice of a Dr. Spock talking to the mother of a child with a fever, or the voice of a Julia Child talking to the cook stalled in mid-recipe, is one of the most reassuring sounds a reader can hear. Always look for ways to make yourself available to the people you're trying to reach.

REWRITING.

Rewriting is the essence of writing well: it's where the game is won or lost. That idea is hard to accept. We all have an emotional equity in our first draft; we can't believe that it wasn't born perfect. But the odds are close to 100 percent that it wasn't. Most writers don't initially say what they want to say, or say it as well as they could. The newly hatched sentence almost always has something wrong with it. It's not clear. It's not logical. It's verbose. It's klunky. It's pretentious. It's boring. It's full of clutter. It's full of clichés. It lacks rhythm. It can be read in several different ways. It doesn't lead out of the previous sentence. It doesn't … The point is that clear writing is the result of a lot of tinkering.

Many people assume that professional writers don't need to rewrite; the words just fall into place. On the contrary, careful writers can't stop fiddling. I've never thought of rewriting as an unfair burden; I'm grateful for every chance to keep improving my work. Writing is like a good watch—it should run smoothly and have no extra parts. Students don't share my love of rewriting. They think of it as punishment: extra homework or extra infield practice. Please—if you're such a student—think of it as a gift. You won't write well until you understand that writing is an evolving *process*, not a finished *product*. Nobody expects you to get it right the first time, or even the second time.

What do I mean by "rewriting"? I don't mean writing one draft and then writing a different second version, and then a third. Most rewriting consists of reshaping and tightening and refining the raw material you wrote on your first try. Much of it consists of making sure you've given the reader a narrative flow he can follow with no trouble from beginning to end. Keep putting yourself in the reader's place. Is there something he should have been told early in the sentence that you put near the end? Does he know when he starts sentence B that you've made a shift—of subject, tense, tone, emphasis—from sentence A?

Let's look at a typical paragraph and imagine that it's the writer's first draft. There's nothing really wrong with it; it's clear and it's grammatical. But it's full of ragged edges: failures of the writer to keep the reader notified of changes in time, place and mood, or to vary and animate the style. What I've done is to add, in bracketed italics after each sentence, some of the thoughts that might occur to an editor taking a first look at this draft. After that you'll find my revised paragraph, which incorporates those corrective thoughts.

There used to be a time when neighbors took care of one another, he remembered. *[Put "he remembered" first to establish reflective tone.]* It no longer seemed to happen that way, however. *[The contrast supplied by "however" must come first. Start with "But." Also establish America locale.]* He wondered if it was because everyone in the modern world was so busy. *[All these sentences are the same length and have the same soporific rhythm; turn this one into a question?]* It occurred to him that people today have so many things to do that they don't have time for old-fashioned friendship. *[Sentence essentially repeats previous sentence; kill it or warm it up with specific detail.]* Things didn't work that way in America in previous eras. *[Reader is still in the present; reverse the sentence to tell him he's now in the past. "America" no longer needed if inserted earlier.]* And he knew that the situation was very different in other countries, as he recalled from the years when he lived in villages in Spain and Italy. *[Reader is still in America. Use a negative transition word to get him to Europe. Sentence is also too flabby. Break it into two sentences?]* It almost seemed to him that as people got richer and built their houses farther apart they isolated themselves from the essentials of life. *[Irony deferred too long. Plant irony early. Sharpen the paradox about richness.]* And there was another thought that troubled him. *[This is the real point of the paragraph; signal the reader that it's important. Avoid weak "there was" construction.]* His friends had deserted him when he needed them most during his recent illness. *[Reshape to end with "most"; the last word is the one that stays in the reader's ear and gives the sentence its punch. Hold sickness for next sentence; it's a separate thought.]* It was almost as if they found him guilty of doing something shameful. *[Introduce sickness here as the reason for the shame. Omit "guilty"; it's implicit.]* He recalled reading somewhere about societies in primitive parts of the world in which sick people were shunned, though he had never heard of any such ritual in America. *[Sentence starts slowly and stays sluggish and dull. Break it into shorter units. Snap off the ironic point.]*

He remembered that neighbors used to take care of one another. But that no longer seemed to happen in America. Was it because everyone was so busy? Were people really so preoccupied with their television sets and their cars and their fitness programs that they had no time for friendship? In previous eras that was never true. Nor was it how families lived in other parts of the world. Even in the poorest villages of Spain and Italy, he recalled, people would drop in with a loaf of bread. An ironic idea struck him: as people got richer they cut themselves off from the richness of life. But what really troubled him was an even more shocking fact. The time when his friends deserted him was the time when he needed them most. By getting sick he almost seemed to have done something shameful. He knew that other societies had a custom of "shunning" people who were very ill. But that ritual only existed in primitive cultures. Or did it?

My revisions aren't the best ones that could be made, or the only ones. They're mainly matters of carpentry: altering the sequence, tightening the flow, sharpening the point. Much could still be done in such areas as cadence, detail and freshness of language. The total construction is equally important. Read your article aloud from beginning to end, always remembering where you left the reader in the previous sentence. You might find you had written two sentences like this:

The tragic hero of the play is Othello. Small and malevolent, Iago feeds his jealous suspicions.

In itself there's nothing wrong with the Iago sentence. But as a sequel to the previous sentence it's very wrong. The name lingering in the reader's ear is Othello; the reader naturally assumes that Othello is small and malevolent.

When you read your writing aloud with these connecting links in mind you'll hear a dismaying number of places where you lost the reader, or confused the reader, or failed to tell him the one fact he needed to know, or told him the same thing twice: the inevitable loose ends of every early draft. What you must do is make an arrangement—one that holds together from start to finish and that moves with economy and warmth.

Learn to enjoy this tidying process. I don't like to write; I like to have written. But I love to rewrite. I especially like to cut: to press the DELETE key and see an unnecessary word or phrase or sentence vanish into the electricity. I like to replace a humdrum word with one that has more precision or color. I like to strengthen the transition between one sentence and another. I like to rephrase a drab sentence to give it a more pleasing rhythm or a more graceful musical line. With every small refinement I feel that I'm coming nearer to where I would like to arrive, and when I finally get there I know it was the rewriting, not the writing, that won the game.

Writing on a Computer.

The computer is God's gift, or technology's gift, to rewriting and reorganizing. It puts your words right in front of your eyes for your instant consideration—and reconsideration; you can play with your sentences until you get them right. The paragraphs and pages will keep rearranging themselves, no matter how much you cut and change, and then your printer will type everything neatly while you go and have a beer. Sweeter music could hardly be sung to writers than the sound of their article being retyped with all its improvements— but not by them.

It's no longer necessary for this book to explain, as earlier editions did, how to operate the wonderful new machine called a word processor that had come into our lives and how to put its wonders to use in writing, rewriting, and organizing. That's now common knowledge. I'll just remind you (if you're still not a believer) that the savings in time and drudgery are enormous. With a computer I sit down to write more willingly than I did when I used a typewriter, especially if I'm facing a complex task of organization, and I finish the task sooner and with far less fatigue. These are crucial gains for a writer: time, output, energy, enjoyment and control.

Trust Your Material.

The longer I work at the craft of writing, the more I realize that there's nothing more interesting than the truth. What people do—and what people say— continues to take me by surprise with its wonderfulness, or its quirkiness, or its drama, or its humor, or its pain. Who could invent all the astonishing things that really happen? I increasingly find myself saying to writers and students, "Trust your material." It seems to be hard advice to follow.

Recently I spent some time as a writing coach at a newspaper in a small American city. I noticed that many reporters had fallen into the habit of trying to make the news more palatable by writing in a feature style. Their leads consisted of a series of snippets that went something like this:

Whoosh!
It was incredible.
Ed Barnes wondered if he was seeing things.
Or maybe it was just spring fever. Funny how April can do that to a guy.
It wasn't as if he hadn't checked his car before leaving the house.
But then again, he didn't remember to tell Linda.
Which was odd, because he always remembered to tell Linda. Ever since they started going together back in junior high.
Was that really 20 years ago?
And now there was also little Scooter to worry about.
Come to think of it, the dog was acting kind of suspicious.

The articles often began on page 1, and I would read as far as "Continued on page 9" and still have no idea of what they were about. Then I would dutifully turn to page 9 and find myself in an interesting story, full of specific details. I'd say to the reporter, "That was a good story when I finally got over here to page 9. Why didn't you put that stuff in the lead?" The reporter would say, "Well, in the lead I was writing color." The assumption is that fact and color are two separate ingredients. They're not; color is organic to the fact. Your job is to present the colorful fact.

In 1988 I wrote a baseball book called *Spring Training.* It combined my lifelong vocation with my lifelong addiction—which is one of the best things that can happen to a writer; people will write better and with more enjoyment if they write about what they care about. I chose spring training as my small corner of the large subject of baseball because it's a time of renewal, both for the players and for the fans. The game is given back to us in its original purity: it's played outside, in the sun, on grass, without organ music, by young men who are almost near enough to touch and whose salaries and grievances are mercifully put aside for six weeks. Above all, it's a time of teaching and learning. I chose the Pittsburgh Pirates as the team I would cover because they trained in an old-time ballpark in Bradenton, Florida, and were a young club just starting to rebuild, with a manager, Jim Leyland, who was committed to teaching.

I didn't want to romanticize the game. I don't like baseball movies that go into slow motion when the batter hits a home run, to notify me that it's a pregnant moment. I *know* that about home runs, especially if they're hit with two out in the bottom of the ninth to win the game. I resolved not to let my writing go into slow motion—not to nudge the reader with significance—or to claim baseball as a metaphor for life, death, middle age, lost youth or a more innocent America. My premise was that baseball is a job—honorable work—and I wanted to know how that job gets taught and learned.

So I went to Jim Leyland and his coaches and I said, "You're a teacher. I'm a teacher. Tell me: How do you teach hitting? How do you teach pitching? How do you teach fielding? How do you teach baserunning? How do you keep these young men *up* for such a brutally long schedule?" All of them responded generously and told me in detail how they do what they do. So did the players and all the other men and women who had information I wanted: umpires, scouts, ticket sellers, local boosters.

One day I climbed up into the stands behind home plate to look for a scout. Spring training is baseball's ultimate talent show, and the camps are infested with laconic men who have spent a lifetime appraising talent. I spotted an empty seat next to a weathered man in his sixties who was using a stopwatch and taking notes. When the inning was over I asked him what he was timing. He said he was Nick Kamzic, Northern Scouting Coordinator of the California Angels, and he was timing runners on the base paths. I asked him what kind of information he was looking for.

"Well, it takes a right-handed batter 4.3 seconds to reach first base," he said, "and a left-handed batter 4.1 or 4.2 seconds. Naturally that varies a little—you've got to take the human element into consideration."

"What do those numbers tell you?" I asked.

"Well, of course the average double play takes 4.3 seconds," he said. He said it as if it was common knowledge. I had never given any thought to the elapsed time of a double play.

"So that means …"

"If you see a player who gets to first base in less than 4.3 seconds you're interested in him."

As a fact that's self-sufficient. There's no need to add a sentence pointing out that 4.3 seconds is remarkably little time to execute a play that involves one batted ball, two thrown balls and three infielders. Given 4.3 seconds, readers can do their own marveling. They will also enjoy being allowed to think for themselves. The reader plays a major role in the act of writing and must be given room to play it. Don't annoy your readers by over-explaining—by telling them something they already know or can figure out. Try not to use words like "surprisingly," "predictably" and "of course," which put a value on a fact before the reader encounters the fact. Trust your material.

Go with Your Interests.

There's no subject you don't have permission to write about. Students often avoid subjects close to their heart—skateboarding, cheerleading, rock music, cars—because they assume that their teachers will regard those topics as "stupid." No area of life is stupid to someone who takes it seriously. If you follow your affections you will write well and will engage your readers.

I've read elegant books on fishing and poker, billiards and rodeos, mountain climbing and giant sea turtles and many other subjects I didn't think I was interested in. Write about your hobbies: cooking, gardening, photography, knitting, antiques, jogging, sailing, scuba diving, tropical birds, tropical fish. Write about your work: teaching, nursing, running a business, running a store. Write about a field you enjoyed in college and always meant to get back to: history, biography, art, archaeology. No subject is too specialized or too quirky if you make an honest connection with it when you write about it.

www.ingramcontent.com/pod-product-compliance
Lightning Source LLC
Chambersburg PA
CBHW080723120726
48001CB00010B/3127